Spikes and Shocks

Angelos Gkanoutas-Leventis

Spikes and Shocks

The Financialisation of the Oil Market from 1980 to the Present Day

Angelos Gkanoutas-Leventis
London, United Kingdom

ISBN 978-1-137-59460-0   ISBN 978-1-137-59461-7 (eBook)
DOI 10.1057/978-1-137-59461-7

Library of Congress Control Number: 2016955694

© The Editor(s) (if applicable) and The Author(s) 2017
The author(s) has/have asserted their right(s) to be identified as the author(s) of this work in accordance with the Copyright, Designs and Patents Act 1988.
This work is subject to copyright. All rights are solely and exclusively licensed by the Publisher, whether the whole or part of the material is concerned, specifically the rights of translation, reprinting, reuse of illustrations, recitation, broadcasting, reproduction on microfilms or in any other physical way, and transmission or information storage and retrieval, electronic adaptation, computer software, or by similar or dissimilar methodology now known or hereafter developed.
The use of general descriptive names, registered names, trademarks, service marks, etc. in this publication does not imply, even in the absence of a specific statement, that such names are exempt from the relevant protective laws and regulations and therefore free for general use.
The publisher, the authors and the editors are safe to assume that the advice and information in this book are believed to be true and accurate at the date of publication. Neither the publisher nor the authors or the editors give a warranty, express or implied, with respect to the material contained herein or for any errors or omissions that may have been made.

Cover illustration: Détail de la Tour Eiffel © nemesis2207/Fotolia.co.uk

Printed on acid-free paper

This Palgrave Macmillan imprint is published by Springer Nature
The registered company is Macmillan Publishers Ltd.
The registered company address is: The Campus, 4 Crinan Street, London, N1 9XW, United Kingdom

Acknowledgements

Writing a book is a long and challenging process, in many different levels. I consider myself very lucky for having the support of a number of exceptional individuals who directly or indirectly helped me in successfully producing a piece of work which I am truly proud of. Therefore, I would like to dedicate this space in recognising those individuals and the support and inspiration that they have so abundantly provided me with.

First, I want to thank Dr. Anastasia Nesvetailova who has been a constant source of support both in my academic and personal life. I appreciate all her patience, time, long discussions and ideas that made my academic experience productive and, most importantly, exciting.

I would also give special thanks to both Daria Nochevnik and Giovanni Mangraviti who took a great interest in my ideas, and through unparalleled patience, attention to detail and creativity have supported me in editing this book, turning it into a beautiful text.

I am lucky to have had a wonderful group of friends whom I would also like to thank for their constant support and influence in my research, who without knowing it gave me the inspiration for studying this topic and subsequently writing this book.

Finally, most of all, I would like to thank my family for all their love and encouragement. For my parents, Ilias and Argyro; and my brother, Petros, who raised me with a drive to challenge and understand the hidden realities in our world, and have always provided me with an environment which supported me in all my pursuits.

Contents

1 Introduction 1
   Reference 9

Part I Conceptualising Financialisation

2 The History of the Oil Market until the 1980s 13
   References 22

3 The Many Faces of Financialisation 23
   3.1 *The Systemic Perspective: Regimes, Capital
      and Financial Geography* 27
   3.2 *The Institutional Perspective: Looking beyond
      the Real Production–Financial Profit Nexus* 31
   3.3 *Could the Institutional and the Systemic Approaches
      be Bridged?* 34
   References 35

4 Finance and the Oil Market: Introducing a Comprehensive
   Approach to Analysing the Financialisation Process 39
   4.1 *Analysing Financialisation: Financial Economy vs Real
      Economy, Conceptualising Performativity
      and the Emergence of New Financial Actors* 42
      4.1.1 *Performativity and Performative Cycles* 44

	4.2 Analysing Financialisation: Behaviour of the Financial Actors	49
	4.2.1 The Behavioural Finance Approach: From the False Sense of Control to 'Mob Psychology'	52
	4.3 Analysing Financialisation: The Role of Technology and Regulatory Evolution	55
	References	60

Part II The Three Phases of Oil Financialisation

5	**Oil Products and Oil-Based Financial Products**	65
	References	68
6	**Oil Shocks as Barometers of the Financialisation Process**	69
	6.1 Oil Shocks and Macroeconomic Performance	70
	References	75
7	**The Three Phases of Oil Financialisation: Early Financialisation (1980–1999)**	77
	7.1 Macroeconomic Background	77
	7.2 The Oil Price Spike Effects	79
	7.3 Financial Dimension of Oil Markets: The Emergence of Oil Futures and Swap Investments	80
	References	83
8	**The Three Phases of Oil Financialisation: Low Financialisation (1999–2002)**	85
	8.1 Macroeconomic Background	85
	8.2 Technology and Regulation: The Three Factors that Changed the Oil Market	89
	References	91
9	**The Three Phases of Oil Financialisation: Advanced Financialisation (2002–2015)**	93
	9.1 Macroeconomic Background	93
	9.2 The 2008 Oil Price Shock and the Role of Speculative Financial Activity	97

9.3	Independent Variables Shaping the Market Dynamics	101
	9.3.1 Drivers behind Speculation on Oil Markets	102
References		105

Part III Financialisation of Oil Market and Evolution of Oil Market's Actor Structure

10 Financialisation of the Oil Market: The Four-Actor Structure — 109

10.1	The New Group of Actors on the Oil Market: Characteristics and Behaviour	112
	10.1.1 Speculative Behaviour of New Market Actors	121
10.2	New Financial Market Players Transforming the Oil Market: Oil Spot Prices and Futures Prices in Focus	125
	10.2.1 Expectations and Behaviour of the Financial Actors	129
References		135

11 Epilogue — 139
References — 148

Index — 151

List of Figures

Fig. 6.1   Recorded oil futures NYMEX against weekly cushing, WTT Spot Price FOB   74
Fig. 10.1 Composition of the recorded oil futures NYMEX   119

List of Tables

Table 10.1	Participants in the 21-day BFOE market and their shares in trading volume	114
Table 10.2	Participants in the crude oil futures market	115
Table 10.3	Correlations between daily price changes of crude oil and other commodities	133
Table 10.4	Correlations between daily price changes of crude oil and oil futures and financial investments	134

CHAPTER 1

Introduction

Abstract This chapter introduces the reader to the imbalances observed in the contemporary history of the oil market. It takes the reader through the arguments which are presented in the book, while introducing theoretical concepts and the historical periodisation which is proposed for the analysis of the modus operandi of the global oil market, as we know it today. It also introduces the argument that instances of speculation, price changes and herd behaviour among oil market participants – which have been increasing over the last two decades – are direct results of the process of financialisation, which has transformed the structure of the international oil trade.

Keywords History of oil market · Global oil market · International oil trade

Wars, embargoes, crises, bubbles, new policy paradigms and the change of the millennium have all left their own mark, one way or another, in the modern politico-economic history book. In a world where international developments have adapted to the speed of the Internet, this book is going to focus on a commodity known to tread very closely to the line where politics meets economics, and is subject to a more 'traditional' set of rules.

Since the emergence of regulated international oil trade – which dates back to 1928 and the establishment of the Seven Sisters (the International

Oil Cartel) – oil has been one of the common denominators shaping and being shaped by international economic and political developments. Its uniqueness is driven by the fact that it is a naturally exhaustible commodity, while at the same time it remains to this day the primary commodity responsible for providing our world with energy. As such, changes to its price have multiple direct and indirect effects on international economies and financial markets and equally, various geopolitical and economic events have been having a direct impact on the oil price.

Over the last decades markets have witnessed a real oil price rollercoaster with West Texas Intermediate (WTI) falling below $30 in 2016 – at the time of the writing – for the first time since 2003. At the end of the last century, on 1 April 1986 WTI fell below $10 per barrel, followed by another plunge at $15.45 per barrel in April 1998 after recovering and reaching a nominal peak of $38 in the beginning of the 1980s (Energy Information Administration [EIA] Open Data). While oil price volatility has been increasing exponentially from the end of the last century, the depth of the latest fall in crude oil prices (2014–2015) was indeed unexpected for a number of stakeholders in the industry.

The current turmoil in the energy sector driven by the tumble in oil prices – as well as the process of rebalancing of the global energy markets against the backdrop of the shale revolution in the USA – reinforces the relevance of the present study. The analysis offered by the author is focused on a specific period within oil market's evolution, which was crucial for defining the modus operandi of the global oil market, as we know it today: the period of financialisation.

Indeed, the two-segment market structure – physical and paper oil – that exists at present is the outcome of the gradual evolution of the oil market, its functioning, constellation of its participants and pricing mechanisms. Upon the emergence of regulated oil trade in 1928, it was a non-competitive market for physical oil with the so-called 'one-base' oil pricing system – that is, real or virtual cost plus (Chevalier 1975) – that later on transformed into the 'two-base' one – that is, real or virtual cost plus pricing for crude oil and netback replacement value for petroleum products (Konoplyanik 2013). While at first the market was dominated by the Seven Sisters, the balance of power between oil corporations and oil-exporting states shifted in the 1970s, and the establishment of the Organisation of the Petroleum Exporting Countries (OPEC) marked the transformation of the respective countries from price-takers into price-makers.

Some thirty years after the market's establishment, in the 1980s the processes of its financialisation had commenced: at that time the model of the oil market started acquiring a much more competitive character, with a new segment of the market – paper oil – emerging along with the penetration of new actors within the market structure. Indeed, from then on it didn't take long before the non-oil speculators from the financial sector had become the key market players and trendsetters, while the oil pricing system was adapted to the rules of the financial markets.

The understanding and the analysis of this period in the global oil market's evolution can arguably provide the necessary clues to interpret the current market dynamics and the relationship between its financial and physical segment, as well as the behaviour of the financial players, while ultimately, shedding light on the interplay between macroeconomic developments over the past thirty years and the oil market.

Cutting through the complex process of oil's transformation from a commodity into a financial asset, this book is intended for experts and academics, as well as the wider audience of energy professionals. Specialists in the field may discover some fresh perspectives on the phenomenon of oil market's financialisation, along with a number of relevant references to the body of literature on the economics of oil and approaches to financialisation as such. At the same time, this analysis is designed for any professional, student or scholar in the field of energy and economics with an interest in the subject matter. The idea behind this work is to enable each curious reader of such kind with an analytical perspective – a lens through which the multi-faced phenomenon of financialisation and its effects on the oil market can be grasped and understood in a comprehensive manner.

To illustrate the traces of the financialisation process that can be found today in numerous market reports, articles and commentaries, one may turn to the recent publication in Bloomberg: "For more than two decades, the people of Vienna have witnessed a peculiar ritual that's been a firm fixture of OPEC's regular gathering: the Saudi oil minister's morning stroll. The walk-and-talk favored by Ali Al-Naimi, Saudi Arabia's top oil official until last month, typically saw dozens of reporters turn out at dawn to accompany him down the city's famous Ringstrasse. His utterances on oil policy often moved the market." (Blas and Mahdi 2016)

Indeed, at present, mere utterances of some leading players on the oil market are enough to trigger the immediate reaction of oil traders, fuelling speculative behaviour; but what are the prerequisites for such price volatility and how are they conditioned by the evolution of the oil market's structure?

These questions are addressed by the hypothesis that the present study sets out to investigate. It first considers whether – and why – the increased price volatility and market speculation of the past three decades can be accounted for by the wider forces of financialisation. Second, it proposes that the topography of today's international oil market was shaped to a large extent by a series of critical events and processes set in motion in the 1980s – the decade that signalled the birth of financialisation in the oil market.

This approach takes issue with conventional accounts of the oil industry that often cite the first oil shock and the rise of the OPEC as evidence that the foundations of the modern oil market structure were laid in the 1970s. The hypothesis of the study proceeds from the recognition that, prior to the 1980s, the oil market was not as accustomed to increases in price levels and market uncertainties, and was alien to such factors as speculative behaviour of market players.

In the present study financialisation is understood as the process whereby financial motives, actors, markets, products and institutions develop in non-financial markets. Conceptually, one may identify two different currents in the academic debate on financialisation. The first strand can be characterised as 'macroeconomic-focused literature', insofar as it conceptualises financialisation as a historical and structural transformation process. The second current of scholarship is more institutional in focus and contemporary in scope. While each of the two theoretical approaches offers insightful analysis of the broader structural dynamics that underpin the complex process of financialisation as well as its effects on the socio-political and economic realms, they fail to provide a platform through which the forces of financialisation can be traced within the very structure of a given market, especially one dependent on physical production, as in the case of commodities.

Drawing on a synthesis of the two sets of academic scholarship on financialisation, as well as on insights from behavioural economics and other social sciences, the present study introduces a two-pillar analytical framework in order to identify the effects of the financialisation process by reference to both macroeconomic dynamics and behavioural practices, as well as to study the effects of this process on markets of oil as a non-financial commodity (initially non-financial).

The above-mentioned framework is then applied to three phases of the financialisation process of the oil market identified by the study, namely: the low phase (1980–1999), the early phase (1999–2002) and the advanced phase (2002–2015). The oil shocks and financial crises that occurred during the respective phases are used as barometers revealing the depth of the financialisation process at a given point in time. The symbiotic relationship between the physical and paper markets is studied through the analysis of the performative cycle that exists between the two market segments.

With the help of this analytical framework, the study unveils the links between the oil market dynamics and macroeconomic performance of a country, focusing on the WTI oil benchmark and the US economy, respectively. Before elaborating any further on the methodology of the study and its structure, it seems necessary to briefly explain the choice of these particular focus points.

In production, contrary to common belief, oil is not purely divided between crude and refined products. As a commodity, oil comes in different qualities, and different variations thereof, based on American Petroleum Institute (API) rating, toxicity levels and sulphur content. The distillate products that result from its processing and refining, such as gasoline, diesel, gas oil, fuel oil, heating oil, naphtha, jet fuel, bitumen and petroleum coke, are equally varied in quality and grade. This variety is crucial for understanding why the economic and political forces within the structure of the oil market lack a universal approach for all oil products. While the equilibrium of demand and supply of crude oil, for example, is highly inelastic, demand alone is sometimes less so, as it depends on the internal – country-specific or region-specific – composition of the actors, as well as their resources, and regulatory and technological potential. A number of benchmarks are used for determining the price level of international crude oils. Among the most widely known are the Light Louisiana Crude (LLS), the Brent and the WTI, which mainly differ in location of trade, quality and calculation method. The choice of price benchmarks varied over time. As stated above, this book focuses on the dominant benchmark in the market between 1980 and 2015, the WTI, which refers to a light, sweet crude oil traded in Crushing, Oklahoma, and marketed in the Chicago Mercantile Exchange. Although the Brent has recently surpassed the WTI as the benchmark of choice in the oil market, its dominance may be short-lived as production levels in the North Sea, where Brent crude oil is extracted, are now dropping. In fact, this usually portends the rise of an alternative benchmark in the oil market. The use of benchmarks as universal references is indeed essential for pricing crude oils produced internationally. The differential between the price of the benchmark and the price of other crude oils is normally calculated based on a combination of different factors, such as chemical properties, refinery yield, quality, acidity, supply-and-demand conditions, producer's commercial strategy, transportation links and costs, as well as geopolitical aspects.

Focusing on a single benchmark allows this study to investigate the rise of a new variable in the structure of the oil market of the past three decades, known as the paper oil market. This refers to markets of WTI and Brent crude oil futures and options traded in the USA and the UK, respectively. One may note that although new markets in refined oil products have appeared in the past few years, mainly in the form of gasoline and diesel futures contracts, they are not yet as popular as their crude oil counterparts.

Having in mind all the above, the choice of focusing primarily on the US economy becomes quite apparent. Not least because the USA rates as a developed economy most directly affected by the oil market by virtue of its status as the leading centre of international finance, exclusive issuer of the international oil currency and, until recently, one of the largest importers of oil. The introduction of horizontal shale-oil drilling technologies dramatically changed the position of the USA on international oil markets, turning USA into a net exporter. At the same time, interpretation of the impact of the new volumes of oil in the USA on the global oil market and the behaviour of market actors today is possible only through the fundamental principles of market functioning. The period between 1980 and 2015 – which the present study is focused on – was crucial for the transformation and establishment of these principles and structure of the oil market, as we know it today.

Turning back to the methodology of the study, it is important to highlight the emphasis that it puts on the role played by human psychology, modes of regulation and fundamental macroeconomic indicators in shaping and magnifying the effects of volatile price levels, speculations – hence in the process of financialisation as such – and, ultimately, in the emergence of financial crises. To that end, in this book the financial crises are always understood to constitute economic crises as well. In this sense, the two-pillar approach adopted by the study – analysing macroeconomic dynamics and behavioural practices – frames crises in financialised commodity markets as outcomes of behavioural trends and actors' expectations on the one hand, and macroeconomic fundamentals on the other.

For the purposes of the analysis in hand, both sets of factors are examined against the background of the types of regulations and technologies that have shaped the market since the 1980s. This twofold approach allows for the study of such factors as the effects of the evolution of the oil market on the macroeconomic performance of the USA. At the same time, it traces these factors to the direct and indirect links between the oil market and the financial sector. Although the development of the oil-based financial sector is analysed as a physically different market to that of oil itself, the increasingly volatile determination of the price of oil is identified as a result of a performative cycle between these two different markets.

This approach is tested against a selection of the most significant financial crises in recent history. Here, crises are employed as methodological tools to better isolate the effects of the financialisation of the oil market. This is because, at these times, the reactions and performance of single market actors are likely to stand out against the norms of the period and abnormalities become easier to spot. The resulting body of empirical evidence, which also includes the period of the 2008 oil shock and the credit crisis, comprises not only the main data and events, but also the motives and dynamics that informed the behaviour of key oil market actors during financial and macroeconomic crises.

As mentioned previously, the past three decades have witnessed a significant inflation of the general price level of oil as well as increased levels of price volatility. The average price level of oil has increased from around $30 per barrel to more than $140 per barrel and has gone down to $30 per barrel over the past three decades and its volatility rate has risen just as dramatically. Traditional approaches to global oil have tended to interpret these developments as the results of fundamental or political factors, such as the scarce and finite supply of world oil reserves or political conflicts in major oil regions. This book looks beyond these conventional explanations by drawing attention to the process of financialisation as the determinant of the oil price hikes and increased market volatility of the past three decades. This period, traditionally described as the era of finance-driven globalisation or financialisation, has witnessed dramatic changes in the dynamics and operations of the oil market. Specifically, financial investments in both the physical and futures oil market have given rise to new forms of market participation. These new actors – mostly financial institutions – have had a profound impact on the fundamental structure of the oil market.

These developments mark a historical shift in a market that was considered a paragon of stability and long-term returns. Traditionally, the price level of oil was dictated by the interaction of demand, supply and mediator forces. Thus, it was influenced only by war, natural disasters, embargoes or politically driven adjustments. As external shocks of this kind were only few and far between, oil prices kept relatively stable. This was the case, for example, before the 1973 oil shock, as well as in the mid-1970s and from 1986 to 1990 and 1991 to 1999. Today, in contrast, expectations of a disturbance in the oil market are enough to influence the futures market and, in turn, to trigger a spiral effect which feeds back into the physical market and thereby affects the price level without any actual change in the underlying fundamental values. It is also worth noting, at this point, that these developments are not unique to the oil sphere, and have been observed in other commodity markets as well.

In fact, the concept of financialisation has been mainly employed in the discussions about financial markets and in the debates about the increasing power and presence of new financial motives, actors, products and institutions in everyday life. Numerous definitions of the phenomenon of financialisation have been proposed over the years, while its interdisciplinary nature is reflected in the fact that contributions to the debate come from a variety of fields. Even so, research in the financialisation of commodity markets began only recently and has so far concentrated primarily on the food market.

This book aims to contribute to this emergent body of scholarship. It will survey approaches to the process of financialisation and then develop a framework to account for its effects on the structure of a 'non-financial' commodity market. While empirically, it presents a fresh analysis of the influence and the effects that the financialisation process have on the structure of the oil market. An important implication of this perspective on the financialisation of commodity markets is that the new configuration of the oil market, shaped as it is by new financial players, technologies and motives, may have played a major role in the genesis of the 2007–2009 international economic crisis.

This research perspective will call for careful empirical analysis of the relevant phases in the evolution of the market that has supplied the primary energy source of all modern economies since the start of the twentieth century – a trend that according to all major international energy forecasting institutions is set to continue into the medium-term future. As a global economic and political resource, oil is central for the well-being and prosperity of individual states and has always boasted pole position on the political–economic agenda of major powers. At the same time, political attention to this resource has escalated in tandem with the complexity of its market functioning. Since the late 1980s, in particular, the oil market has grown more elaborate and multi-layered in terms of both real commodity production and financial operations.

The findings of the present study demonstrate that instances of speculation, price changes and herd behaviour among oil market participants – which have been increasing over the last two decades – result precisely from the process of financialisation, which has transformed the structure of the international oil trade.

Reference

Blas, J., & Mahdi, W. (2016, May 31). Saudi's new top oil official brings fresh style to OPEC. Bloomberg.

Chevalier, J. (1975). *The new oil stakes*, translated By Ian Rock. MI: University of Michigan.

Konoplyanik, A. (2013). Global oil market developments and their consequences for Russia. In A. Goldthau, *The handbook of global energy policy*. West Sussex, UK: Wiley-Blackwell.

PART I

Conceptualising Financialisation

CHAPTER 2

The History of the Oil Market until the 1980s

Abstract This chapter provides a brief overview of the history of the oil market, from its early beginnings to the 1980s. An understanding of the historic background is essential to study its financialisation, which is categorised into three periods. Topics such as peak oil, the creation of OPEC, the abolition of the gold standard and the 1970 crises are also outlined before introducing to the reader the concept of financialisation and the principles which drive the periodisation applied throughout this book.

Keywords Historic finance · Oil market · Gold standard · Peak oil

Compared to other markets, the characteristics of the oil market are especially complex. As a non-renewable resource, oil is not only naturally scarce, but also exhausted through use. Therefore, the general expectation is that, at some point in the future, it will deplete. Herein lies the special complexity of this market: it is as yet impossible to know exactly how much oil is still available. Even though much of the land has been explored, the same cannot be said for undersea fields. Moreover, oil-producing countries have not provided – and cannot provide – reliable statistics on the amount of oil that is potentially available in their territory, not just on political grounds but also because "reserves are 'inventories, constantly used up and replaced' [and] estimates of 'total production' from a reservoir over time" (Labban 2010). This leaves any estimate of the international availability of oil open to question.

Be that as it may, global dependency on oil as primary energy source is sufficient proof of the central role that the oil market plays in the modern world. From the production of plastic to that of the fuel that powers cars, airplanes and ships, oil has become an integral part of everyday life. Thanks to its relative abundance, ease of access and low costs, the ubiquity of oil, especially in the West, is now taken for granted. As a result, demand for energy has become infinitely inelastic with the advance of technology: as long as it remains the most widely used and preferred energy source, oil will retain monopoly control of this sector.

This monopoly is easily explained by the fact that to date no other source of energy is able to compete directly with oil, particularly when it comes to features such as those of its flexibility, affordability, availability and range of applications. Energy sources, such as solar, hydro, wind and even natural gas, are not yet able to outperform oil in all these fronts. It is worth noting, however, that the reasons behind this status quo run deeper, and are partly political–economic in nature. Indeed, the study of technology and the oil markets reveals an interesting pattern, whereby research and technological breakthroughs in alternative energy sources mostly occur during periods of high oil prices (Adelman 2004). In other words, people and governments seek ways to replace oil only when its price becomes too high for them to sustain. Monopoly control of the energy sector is therefore highly contingent upon prices of oil remaining at a level that deters technological advances in alternative energy sources.

The history of this successful commodity market, in its current form, dates back to 1862, when Rockefeller made the first investment in oil. This resulted in the creation of Standard Oil, which had complete monopoly in transporting and marketing oil. This monopoly persisted up until 1901 when the Spindletop field in Texas, the largest oil reserve in the American continent, was discovered. Andrew Mellon, a banker and industrialist, raised the capital to form the Gulf Oil Corporation, the first company to exploit this new reserve. A number of other corporations followed suit. Among them, the Texas Corporation (Texaco) and the Shell Transport and Trading Company of London were the largest.

A number of oil discoveries around the world, especially in Russia and India, created new competition in the world markets for Standard Oil; at the same time, the merger between the Royal Dutch and the Shell Transport and Trade Company gave birth to the Shell group. However, the biggest hit came with the US antitrust legislation in 1911. The Standard Oil monopoly was broken and the company was divided, paving the way for the first oil cartel, later named 'Seven Sisters'. The Seven Sisters consisted of BP, also known as Anglo-Persian Oil Company; Gulf Oil; Standard Oil of California; Texaco; Royal Dutch Shell; Standard Oil of New Jersey; and Standard Oil Company of New York. The Seven Sisters, along with the other large international oil companies that later entered the market, were known as the 'majors'.

Over the following decades, the major oil companies, with national government backing, engaged in an international competition for control over reserves in the Middle East. The majors tried – sometimes with questionable approaches – to reach concession agreements with the governments of oil-producing countries. In 1949, the majors controlled 55 per cent of the crude oil production in non-communist countries; the figures of refining, transport and marketing were not far off. Most of these concession agreements required companies to pay oil-producing governments a tax on profits from oil exports. This tax was calculated using a public price, known as the posted price, which accounted for 50 per cent of the profits.

Throughout the 1950s, new oil discoveries and the rapid expansion of supply led to continuous reductions in the market price of oil, which led oil-producing countries to request frequent readjustments in the posted price. The reduction on tax payments to oil-producing countries eventually triggered the creation of one of the most well-known international organisations, the OPEC. Formed in 1960, the OPEC initially comprised Iran, Iraq, Kuwait, Saudi Arabia and Venezuela, although its membership reached as many as thirteen by 1975. The main goal of the OPEC at the time was to keep posted prices as high as possible. In the 1960s, the organisation secured small concessions and redefined the posted price from a market-driven figure into a tax reference price. However, the OPEC was ill-equipped to act as a cartel, owing to the diversity among its members. So, until the 1970s, the majors were able to defend their best interests vis-à-vis the governments of oil-producing countries, thereby maintaining their favourable market position.

In 1970–1971, Libya's successful pressures for progressive production cuts from Occidental, an American independent oil company, marked the OPEC's first victory. As Occidental was dependent only on Libya's oil, and Exxon withheld assistance, the company had no choice but to abide by Libya's demands (Bromley 1991). This sent a wave of gradual increases in crude oil posted prices from Libya to Nigeria and all the Mediterranean ports that controlled the traffic of Iraqi oil. The initial 30 per cent increase in posted price and 4–8 per cent increase in tax rate was swiftly reproduced by the rest of the OPEC. The closing of the Suez Canal, the damage in the Trans-Arabian oil pipeline and the 1967 Arab–Israeli War also played a significant role in the above price increases (Odell 1986).

In 1971, six of the OPEC countries allegedly entered a bargaining process with two major oil companies. Nevertheless, not much bargaining is believed to have taken place at these meetings as both the US government and the majors had strong economic and security interests in an oil price increase. That year, the Middle Eastern members of the OPEC increased the price of crude oil in fulfilment of what are now known as the Tehran and Tripoli agreements. Though the general conditions of the market barely changed as a result, the signing of these agreements marked the first time that the OPEC countries negotiated as a unit with the majors and, even more strikingly, that they did so with the approval of the US State Department. Unbeknownst to both the majors and the US government, this was the beginning of the end for their control over the oil industry.

As described thus far, the history of the oil market up to the 1970s features as many as three distinct actors, namely oil-producing countries, oil-importing countries and oil corporations, whose responsibilities ranged from extraction to marketing. During the first years of the oil market, the balance of power between these three actors was widely in favour of the oil-consuming countries. By pulling the strings of the majors, oil-consuming countries commanded a cheap and secure supply of oil, while also overseeing the mechanisms and profitability of extraction and marketisation. It is not surprising, therefore, that the main economic powers of the time, such as the USA, the UK, France and the Netherlands, went to great lengths to secure exclusive access to the supplies of several oil-producing countries for their respective oil corporations as a way to pursue their geopolitical agendas (Tanzer 1974). In contrast, oil-producing countries were the least influential actors in that period, owing to the lack of cooperation and leadership. As a result, they gained minimal returns on the volumes of oil extracted until then.

In 1971, six of the OPEC countries allegedly got into a bargaining process with the two major oil companies; however, the consensus is that not much bargaining took place during these meetings. According to Simon Bromley (1991), the oil multinationals were driven by the US government which had clear interests in an increase in the oil price level. He points out that a series of challenges against the US global sovereign status, which took place almost simultaneously during that period, led the State Department to take up this opportunity and pursue this policy.

One such challenge for the USA was the reduction in its share of the world's oil production, which occurred for the first time. This was mainly due to the discoveries of new reserves in the Middle East during the 1960s which reduced the market price of oil. As a consequence, many of the US wells became commercially unsustainable due to the low quality of the oil that they produced and high refining costs. An increase in the price of oil at this point in time would resurrect the oil production industry of the USA and reduce imports.

He raises the issue of doubts over the US power and authority after the unexpected defeat in Vietnam, and proposes that an increased price of crude oil would resurrect the US influence in the Middle East as the oil-backed regimes of Saudi Arabia and Iran would be strengthened, and the petrodollars that they would acquire would find their way back to the USA through an increase in demand for US military products.

Next, he points out the fact that the USA was preoccupied with the major threats of the Soviet Union which was continuously expanding its military power, as well as the West European countries and Japan which were rising economic powers and had just started to approach the Middle Eastern countries in order to promote their interests. Hence, he suggests that an increase in the price of crude oil would be a major blow for the main of all these emerging powers since they were heavily dependent on oil.

Thus, he argues that when the Nixon–Kissinger team saw an opportunity to re-establish the USA in its role as the leader of the West, they were keen to seize it, and consequently, along with the 'majors', they allowed the OPEC members of the Middle East to increase the price of crude oil in 1971 through what are now known as the Tehran and Tripoli agreements. The situation did not change much immediately after these agreements, but what had changed though is that for the first time the OPEC countries, with the 'blessing' of the US State Department, managed to negotiate as a unit with the 'majors' and signed the Tripoli and Tehran agreements. Neither the 'majors', nor the US government did at any point realise that this was the beginning of the end for their control of the oil industry.

Things did not turn out as planned though, and when President Nixon took office, he was faced with a very tough situation as from one side the vast expenditure required for the Vietnam War along with the increasing oil prices was putting the US economy in a worsening economic position. While the economy was in a recession, the Federal Reserve was unable to boost the economy in a timely fashion as long as the USA remained a part of the Bretton Woods system, and President Nixon was the target of all the critics of the economy.

Action was required and pressure was put on President Nixon as he was losing the trust even from people of his own political party. Therefore, on 15 August 1971, as described by Joel Kurtzman,

> Action was what they got. To fix the 'sick' economy, Dr Nixon, as he was called in a subsequent headline in the New York Times, tried shock therapy. In a televised speech Nixon, upper lip wet with sweat, voice resonant, announced that he had signed a presidential order freezing wages and prices for ninety days. He said he would try to persuade the Congress to make it illegal for unions to strike during that time, that he imposed 10 per cent surtax on imported automobiles and other products, and that he would propose a cut in income taxes to the Congress. He also said, to quote the day's vernacular, that he had closed the 'gold window'. (Kurtzman 1993, p. 51)

The last announcement was to create the biggest change in the international economic system since the Great Depression. What it meant was that it effectively cut any official link that existed between the dollar and gold. In fact, the USA refused to sell gold to foreign banks even at this increased price. A devaluation of the dollar took place shortly, with the aim of boosting productivity financing the ever-increasing budget deficit.

The year 1973 also marked the beginning of the Arab–Israeli War which destabilised the Middle East. Cut backs in the production of crude oil took place immediately, creating a supply shortage. This restriction of supply created pressures on the price of crude oil; pressures were intensified when Saudi Arabia decided to use its control over oil production as a political weapon. A 10 per cent reduction in the production of oil in Saudi Arabia was introduced along with an embargo on oil exports to the USA and the Netherlands. The shortage created by the Arab–Israeli War, along with the Saudi Arabian policies, increased the price of oil very rapidly. The rest of the Arab oil-producing countries, with the exception of Iraq, followed the Saudi Arabian example and joined in the embargo.

The non-Arab oil producers did not join their Arab counterparts; on the other hand, they did take advantage of the situation by increasing the prices of their crude oil. This embargo lasted for nine months which put the USA on the verge of an energy crisis as its domestic oil production was not nearly enough to satisfy the demand, while any oil reserves were consumed during the previous years. This embargo lasted for a number of months, ending in January 1974; but the prices were never readjusted to their previous level.

The transformation that took place in 1973 was that what began as an issue between oil companies and oil-producing countries, turned into an international political and economic struggle with hegemonic influences. The balance of power within the oil market changed, and the OPEC countries realised that they could control the 'majors', and consequently, the nations to which the 'majors' belonged by managing the oil supply. The demand for energy has been ever increasing, while the alternative energy sources of the time were dying out. The OPEC countries realised that if they were to act as a unit they would not have to settle anymore for the small tax payments over the posted price paid to them by the 'majors'. Instead, they could increase the price of oil by playing with the simple economics of supply and demand and enjoy supernormal profits. This led Pierre Terzian to argue that the King of Saudi Arabia, "Fiasal did probably more damage to the West than any other single man since Adolf Hitler" (Terzian 1985, p. 201). For the following years the situation would not change significantly, with the highlight being the secret meeting between the USA and Saudi Arabia in 1974 where the US dollar was established as the official currency for oil trading.

The second oil shock which took place towards the end of the same decade was mainly driven by events in Iran. The Shah wanted to drive up the prices of crude and therefore he attempted to introduce cutbacks in production. By 1978, the Iranian oil production had dropped to its lowest levels for twenty seven years. The conflicting domestic and international interests, however, triggered a revolution in 1979, which led to the deposition of the Shah. Nevertheless, the USA–Iranian relationships had not normalised after these events, as in 1979 Iran took US citizens hostages while the next year the USA halted all imports from Iran. The damage in the oil market was done.

This oil shock, even though very extensive, was not as harmful for the US economy as the first one. The focus started to shift gradually away from the USA as a series of events was completely altering the financial scenery in Western Europe. While the European countries were victims of the US interest-driven policies in the oil market which increased the oil prices, as most of them did not have any alternative or domestic production of oil, vast inflows of dollars were flooding their economies. During the first and second oil shock, the Middle East was accumulating enormous amounts of dollars especially since the agreement of 1974 when the dollar became the official currency for transactions in the oil market.

The increased oil prices since 1970 raised dramatically the dollar holdings of the oil-producing countries, while the agreement of 1974 made sure that these holdings were only in dollars. The global oil transactions were taking place in dollars, a fact that increased the international demand for the US currency.

However, the increased stock of dollars which was now being accumulated in the Middle East started to unsettle international financial communities, while the percentage of crude oil marketed directly by OPEC governments into the world markets increased by almost 500 per cent during this decade (Luke 1983). The fact that these inflows were triggered by the increased oil prices proposes why they were later named as petrodollars. This immense increase in the current account balances of the OPEC countries led to rapid growth in their GDP levels. The action of OPEC 'was undoubtedly the greatest forced redistribution of wealth in the history of the world' (Glipin 1981, p. 208).

This balance of power in the oil market turned on its head over the 1970s. The foundation of the OPEC, and the 1973 oil shock, in particular, were instrumental in bringing the balance of power to the point where oil-producing countries gained full control of the market, while oil-importing countries were left as price-takers. Deprived of market power, the once powerful majors gradually turned into transnational corporations in order to promote their own interests and fight for their own survival. At the same time, oil-producing countries launched a new type of Oil Company, that is, national oil corporations, in order to better defend their newly acquired market power.

This understanding of the structure and actors of the oil market until the 1980s is central to the argument developed in this book. This is because the triangular structure of oil-producing countries, oil-importing countries and oil corporations underwent a fundamental transformation since the 1990s. With the emergence of the process of financialisation in the dynamics of the oil market, a fourth group of actors wended its way in the already delicate structure of the oil market. As argued in the following chapters, this reconfiguration had a transformative impact on the very processes, functions and dynamics of the market described thus far.

It is worth mentioning that the periods of oil market's evolution, identified in the present study, are broadly in line with a number of other reputable classifications of oil market's development stages. Konoplyanik (2013) proposes to distinguish between five periods of oil market development, namely: (1) 1928–1927 – the first period when only the physical oil trade existed, while the market was non-competitive and dominated by International Oil Cartel, with the one-base system pricing and traditional long-term concession agreements prevailing; (2) 1947–1969/1973 – the period when the model of the oil market was still non-competitive, with the Oil Cartel dominating the market, however, then the two-base pricing system emerged and the concession agreements became to modernise, the production sharing agreements (PSAs) were being introduced; then the author highlights the transition period from the monopoly of the Seven Sisters to that of the OPEC; (3) the third period – while still being characterised by the non-competitive market character and the dominating position of the OPEC – is marked by the changes in the pricing system. Contractual and spot oil prices were introduced and OPEC was determining the so-called official selling prices (both cost plus and net forward) with contractual structures being linked to spot quotations; another transition period takes place between 1985 and 1986 marking the shift from net-forward to net-back crude pricing that eventually becomes based on the futures quotations established at the main petroleum exchanges and market places; (4) 1986–2000 (2004) – the fourth period is marked by the emergence of the second segment of the oil market – that of paper oil. As Konoplyanik maintains, at that period oil prices were established mostly at the market places and were largely driven by oil hedgers (the net-back from futures oil quotations formula established at the previous period remained). At that period, the foundations of the global oil market and its institutes were laid, on the basis of the modus operandi of the established institutes of the financial markets. Having said that, at the time, market fundamentals still constituted the defining factors for oil pricing. (5) Mid-2000s till present – Konoplyanik argues that this last period is characterised by the dominance of the paper oil market in volumes of trade, and this segment of the market indeed becomes an important element of the global financial market. The global institutions for paper oil market took shape and non-oil speculators eventually become the key market players, hence oil pricing is being established largely outside oil market places. The technological advancement and the functioning of the paper oil markets allow fundamentally new players to penetrate it. While the net-back from futures oil quotations remains, the market is flooded by oil-related financial derivatives. The latter became the key factor defining the price of oil.

This periodisation is undoubtedly relevant for the study at hand, while the convergence of views given the independent analysis of the author and the respective group of scholars creates a notable case in point. However, the present study focuses specifically on the period between 1980s and 2000, and the oil market financialisation process, taking place at that time. Hence, this book introduces a more detailed breakdown of the phases within the oil market financialisation period, that is the low phase (1980–1999), the early phase (1999–2002) and the advanced phase (2002–2015), focusing on the changing constellation of actors within the three-actor structure of the oil market that existed pre-1980s and the emergence of the paper oil market segment.

At the same time, as described in the introductory chapter, this study places its focus specifically on the development of the WTI benchmark and the interplay between the macroeconomic performance of the USA and the oil market dynamics, employing oil shocks as barometers of the oil financialisation process.

References

Adelman, A. (2004). Is the oil shortage real? Oil companies as OPEC tax-collectors. *Foreign Policy*, 9, 69–107.
Bromley, S. (1991). *American hegemony and world oil*. University Park: Pennsylvania University Press.
Glipin, R. (1981). *War and change in the world politics*. Cambridge: Cambridge University Press.
Konoplyanik, A. (2013). Global oil market developments and their consequences for Russia. In A. Goldthau (Ed.), *The handbook of global energy policy*. West Sussex, UK: Wiley-Blackwell.
Kurtzman, J. (1993). *The death of money*. London: Little, Brown and Company.
Labban, M. (2010). Oil in parallax: Scarcity, markets, and the financialization of accumulation. *Geoforum*, 41(4), 541–552.
Luke, T. (1983). Dependent development and the Arab OPEC states. *The Journal of Politics*, 45(4), 979–1003.
Odell, P. (1986). *Oil and world power* (8th edn). London: Penguin Books.
Tanzer, M. (1974). *The energy crisis, world struggle for power and wealth*. London: Monthly Review Press.
Terzian, P. (1985). *OPEC: The inside story*. London: ZED Press.

CHAPTER 3

The Many Faces of Financialisation

Abstract This chapter acts as an extensive literature review on the theory of financialisation, examining its origins, the plethora of different definitions which have been developed through time, as well as the strengths and weaknesses which can be attributed to its two main schools of thought. Performing a close examination of the arguments made by the authors who have argued for both the institutional and systemic approaches to financialisation, this chapter proposes that a bridge can be reached between the two in forming a theoretical framework which can be applied in the study of the effect of financialisation in a commodities' market.

Keywords Financialisation · Definition financialisation · Commodities' market

The past thirty to forty years have witnessed an unusual development in the macroeconomic landscape of a number of advanced economies: the latter, at times of dwindling physical investments, reported sustained increases in profit rates. In response to the seemingly counter-intuitive nature of this phenomenon, many analysts have been left scrambling for a coherent explanation.

How can the nonlinear relationship between the real production and financial profit be grasped and explained? Is it a consequence of the evolution of the capital regime or some specific sociopolitical processes within it? What forces drive the process of decoupling of real production from financial profit? And how do the non-financial markets ultimately engage in the financial sector? These are only few crucial questions that emerge in the context of the new macroeconomic phenomena that require a comprehensive explanatory paradigm. The concept of financialisation, however recent or controversial, promises to offer precisely that: a paradigm through which one can understand the complex relationship between the real production and financial profit and analyse the processes that lie behind it. That is, the kind of evolutions and transformations that occur within the structure of the financial sector as well as between finance and other areas of human activity.

This is a broad understanding of financialisation, which resonates with the definition proposed by Stockhammer (2004). In fact, as an academic term, the notion of financialisation has been part of the social–scientific vocabulary for over twenty years and can hardly be narrowed down to a single interpretation. Up until now, no less than fifty different definitions have been provided by scholars from a wide variety of academic disciplines, including human geography, sociology, political science, economics and political economy (Nesvetailova 2007).

In what follows, we will take a step aside from the oil market focus and concentrate on the notion of financialisation as such, only to return back to the crux of this book's argument with a sophisticated analytical tool – conceptualised and articulated according to the academic tradition that it is part of.

Before unfolding the spectrum of key approaches to financialisation, it is worth noting that there is a golden thread running through most of them – a common understanding of the origins of this phenomenon. While disagreements abound on the nature, originality, significance and consequences of the process of financialisation, it is generally agreed that the last forty years have witnessed the growing importance of "financial motives, financial markets, financial actors, and financial institutions in the operation of the domestic and international economies" (Epstein 2005). The roots of this form of financial ascendancy are considered to coincide with the announcement of the end of the gold standard and the subsequent demise of the Bretton Woods system, which marked the beginning of what is known as the post-Fordist period (Jorda et al. 2011). This was the biggest watershed in the international economic system since the Great Depression: the official link hitherto existing between gold and the US dollar, with the former acting as the underlying commodity of the latter, ceased to apply and the US Federal Reserve, now able to print money at will, was freed from the obligation to sell gold to foreign central banks in exchange for dollars.

This historical juncture in the international monetary system was central to shaping the financial status quo of the twenty-first century. As Harry Magdoff and Paul Sweezy (1969) maintain, the abolition of the gold standard and the introduction of the 'paper dollar standard' was a fatal blow to the international monetary system. This is because the USA traded the privileges of controlling the gold exchange standard with the even greater privileges that came with control of the paper dollar exchange standard.

Beyond that, with the abolition of the gold standard and the elections of Ronald Reagan and Margaret Thatcher, the financial markets both in the USA and the UK underwent extensive deregulation policies. The Big Bang deregulation of the City of London in 1986 in the UK and the abolition of the Glass-Steagall Act by President Bill Clinton in 1999 in the USA are classic examples of this historical development. Finance had now "penetrated across all commercial relations to an unprecedented direct extent" (Fine 2009) and transformed the functions of everyday life as never before.

The role of financialisation in the evolution of the contemporary economic, social and political systems is indeed unquestionable, but what lens should one choose when attempting to analyse the phenomenon of financialisation as such?

It is worth mentioning that there is a pleiad of authors who employed the ideas, which only later came to fall under the rubric of financialisation. Most of them attempted to highlight the growing influence of capital-market over bank-based financial systems. Years before the concept of financialisation was introduced, Hilferding used similar concepts to describe the reallocation of political and economic powers in the social class structures. The crux of Hilferding's (1910) argument is that the rise of financial capital is responsible for the transformation of the modern capitalist system on the grounds that the industrial and banking capital increased its dependence on financial investment. This, in turn, allows the financial sector to restructure the economy to its own advantage. The interconnection of finance and industry is explained in terms of interlocking appointments, exchange of information and joint decision-making, made possible by the fact that financial and industrial capital share common interests in the profitability of their financial endeavours. Hilferding also studied the idea of imperialism as an economic rather than a political process, and related this to the idea that large monopolies have a propensity to depend on bank and financial capital. In his view, this form of economic imperialism was directly responsible for the introduction of trade barriers, export of capital and militarism. In fact, Lenin was adopting Hilferding's main ideas when he proposed the definitive Marxist theory of imperialism based on the concept of 'parasitical rentiers' (Lapavitsas 2009), and even though Hilferding's approach is considered outdated and overly centred on the Austrian and German economies, his ideas have formed the theoretical basis of a large part of the literature on financialisation.

Among other accounts of financialisation – falling under the regulation theory umbrella – is that the one of Robert Boyer (2013), notable for its attempt to combine the analytical frameworks put forward by a number of prominent authors, namely Minsky, Hayek, Fisher and Keynes. Through the study of the 2008 global financial crisis of 2008, Boyer analyses the impact of excess financialisation on economic sustainability of any financial regime. He argues that – contrary to the common theories suggesting the benefits of financialisation for the real economy, mainly when it comes to efficient capitol allocation – the process of financialisation can endanger the real economy once financial innovation becomes out of hand. Hence, the argument of Boyer for a greater social control of the respective innovation dynamics being the only way to tackle the uncertainties and the instability brought about by the unleashed financialisation process. As Bruce G. Carruthers (2015) argues in turn that, as the financialisation processes were picking up steam, "the absence of public regulation bolstered private control over the market and was justified on the grounds that sophisticated private risk management (using credit ratings and value-at-risk models) would combine with self-interest to ensure market stability". Carruthers maintains that the phenomenon of financialisation gained its momentum and penetrated the markets due to the flexibility of the underlying institutions of the capitalist systems, putting forward the role of institutional changes in this process.

All in all, the contemporary academic tradition identifies two main approaches to financialisation: the historical one, in the form of a systemic analysis of 'finance-led capitalism', and the institutional one – a more contemporary perspective. While the former approach focuses on the effects of the growing influence of the financial industry on the broader economy, the latter draws special attention to how the "various realms of human activity are absorbed by the financial dynamics and become new elements of the financial system" (Nesvetailova 2007). In what follows, some key perspectives from both of the approaches to financialisation will be reviewed, unveiling the multi-faced nature of this concept. While the next question that this study will pose would be whether and to what extent those two approaches could be bridged?

3.1 The Systemic Perspective: Regimes, Capital and Financial Geography

Systemic approaches to the study of financialisation generally focus on the rising influence of finance in the national and international economic systems over the course of the second half of the twentieth century and advance the argument that the process of financialisation has led to the progressing decoupling between real production and financial profits at the national and international economic level.

Among the earliest systemic analyses is the work of Baran and Sweezy (1966), where the process of financialisation is described as resulting from the transformation of the regimes of capitalist accumulation. This is ascribed to the fact that, in mature capitalist societies, production is unable to fully absorb new investment from established monopolies and, as a result, investment capital is induced to flood into speculative financial markets. On a similar note, Foster (2008), in his study of the correlation between financialisation and the deceleration of production investment, concludes that it is the very monopoly stage of capitalism, as described by Baran and Sweezy, which is responsible for creating demand for novel financial products. In fact, he later goes on to claim that, at such an advanced stage of capitalism, economic activity might altogether be shifting from real production to finance. Other notable systemic accounts are found in Lapavitsas (2010), who describes the general increase in financial profits as the outcome of the expropriation of workers' income on the circulation sphere, which, in disagreement with the latter, points to the wider structural transformations triggered by the growth of interest-bearing capital throughout the institutional structures of the capitalist system.

Another point of departure for defining the process of financialisation in systemic terms would be to start from the prior definition of 'capital'. This is the approach suggested by Paulani (2009), as the concept of self-expanding capital fails to incorporate the crucial character of capital as an abstract, centrifugal form from which its own content tends to escape. In other words, this approach establishes that capital is more than self-expanding in character, as it tends to become autonomised in its social form. Therefore, in the Marxian understanding of financialisation, the autonomisation of capital should be approached as the natural inclination of the social forms of the capitalist system to detach themselves from their own base.

The 1980s crisis, considered as a watershed for the development of the financial markets, has been serving as an important case study for a number of scholars seeking to explain the nature of the financialisation process. Bob Rowthorn (1980) argues that this crisis was an inevitable consequence of the flawed structure of the capitalist system, whereby overproduction leads to redistribution and social deconstruction. At the same time, Rowthorn incorporates in his approach the role of the banking industry, especially the operation and regulation thereof. When examining the crisis from a more political perspective, he puts forward the idea that "the imposition of credit restrictions and the adoption of monetary targets caused the rate of profit to fall and this in turn led to a generalized world recession. The system [...] destroyed faith in the market by rewarding the strong at the expense of the weak" (Strange 1997).

Rowthorn's approach has some common features with that adopted by Samir Amin (1980), who focuses on the relationship between international capital, national governments and developing countries. Amin illustrates how the costs of borrowed capital, technology and falling profit rates shift from multinationals and national governments to developing nations as a consequence of cheap labour and primary goods trade. In contrast, Ricardo Parboni (1980) concentrates on the role of the international monetary system and argues that currency policies played a major role in isolating the USA from the worst effects of the world depression until 1979.

A significant contribution to the systemic debate on financialisation was also made by Giovanni Arrighi (1994). In his discussion of Marx's general formula of capital, Arrighi maintains that commodity growth, when achieved by investing money into production, eventually leads to money growth as money breaks away from commodity production. He effectively introduces the world systems approach to the study of financialisation, by arguing that the financial and money flows flourish when real production stagnates. More specifically, the systemic cycle of accumulation implies an over-accumulation of capital that results in financial expansion, as investments in production growth are not as efficient as investments in the financial sphere. Arrighi places financialisation within a cyclical framework of the international economy, where hegemonic capitalist formations evolve and succeed each other in a cyclical pattern. According to Arrighi, financialisation corresponds to the autumn of the hegemonic power's lifetime, as the financial sector takes over the productive one. He proposed that financialisation, as a political–economic term, can be used to refer to the "'prolonged split between the divergent real and financial economies' and the defining moment of international hegemonic transition" (Nesvetailova 2007).

This approach conceptualises financialisation as an outcome of the over-accumulation of the current capital regime, which facilitates the movement and extensive relocation of capital across geographical spaces, in a way that allows for the creation of bubbles in some places and production-focused investment in others. The globalisation of the financial markets and the resulting commodification of money have precipitated a global deindustrialisation trend; it has also intensified the social cleavages created by money, being as it is unencumbered by physical, fiscal or economic barriers. The rules imposed by the globalisation of the financial and money markets have engendered new processes of diversification in contemporary societies under the pressure of international competitive interests and the 'ubiquitous money fetish' (Altvater 1997). Commenting on this point, Palley (2007) adds that income and wealth inequalities are, in fact, a predictable result of the process of financialisation, insofar as the latter entails a disconnection between productivity growth and wages.

On a similar note, David Harvey (1982) argues that financial geography is key to understanding the way capital tends to elude crises by moving into new spatial and institutional locations and causing, in its wake, new strands of capital to collide, collude or compete. To that end, a notable financial–geographical approach to financialisation is offered by Pike and Pollard who argue that the process of financialisation is "broadening and deepening the array of agents, relations, and sites that require consideration in economic geography and is generating tensions between territorial and relational spatialities of geographic differentiation" (2010). This conceptualisation is based on the idea that the relational space between these actors has evolved in tandem with the development of the new social networks and patterns of actor interaction. It is this form of geographical evolution that, in turn, changes the allocation of economic resources and the shape of the economic landscape to accommodate the new financial practices. Overall, this geographical approach engages with the growing social, spatial and political reach of financialisation in relation to three main concepts: increased risk, uncertainty and volatility (Pike and Pollard 2010).

More recently, Smart and Lee (2003) have conceptualised financialisation as a political process that acts by shaping and reshaping the relationship among the actors and collectives that operate within the structure of the international economic system. They suggest that financialisation contributes to the shift of resources by reshaping social relations on a variety of different levels, from production and consumption to state and society. In other words, this analytical approach to the study of financialisation, unlike those based on the idea of over-accumulation, requires in-depth analysis of the everyday, strategic terrain in which financial activities take place.

The increasing importance of 'shareholder value' in the mode of modern Western capitalism is placed in focus of a number of contemporary studies on the topic. The rationale behind such type of an approach suggests that firms tend to develop an increased preference for financially driven short-term profitability at the expense of investment and real production growth. Hence, these studies focus on firms as "the key agents of adjustment [...] whose activities aggregate into overall levels of economic performance" (Hall and Soskice 2001, p. 6). This view, for example, is corroborated by the findings of a study by Lazonick and O'Sullivan (2000), who report a strong connection linking the rise of neoliberal regimes with the rise of shareholder value and the downscaling of firms.

Similarly, political economists approach the yawning gap between financial profits and real-production profits as the decoupling of the financial economy from the real economy. This decoupling is argued to stem from the following process: with financialisation "profits accrue primarily through financial channels rather than through trade and commodity production" (Krippner 2005, p. 174) and consequently, many firms are compelled to intensify their financial operations in order to maximise their profits. Building on Arrighi, Krippner has argued that firm revenues can in fact be used as a reliable indicator of financial expansion, a practise that has come to be known as 'short-term performance obsession' (Rappaport 2005).

For all their different takes on the causes and nature of financialisation, all systemic approaches can be argued to centre on the dimension of society and on the structural consequences of the rise of financial profits. Here, they all identify a gap between the traditional value-production processes of real production and financial profit creation. The structural changes caused by the process of financialisation are considered to be spatial, institutional and societal. Finally, all systemic approaches tend to consider the shift from real production value to financial profit creation responsible for the creation of new dynamics at the national and international level.

These approaches to financialisation, however, tend to be limited by their level of generality. Such accounts are typically long on the dynamics and effects of the gap between real production value and financial profits at the national and international levels, but short on the actual mechanism of said relationship. The decoupling between the rise of financial volume creation and real production is generally assumed to generate two autonomous entities, driven by different elites and dynamics. In turn – the argument goes – these new elites and dynamics change the national and international status quo in terms of value creation, balance of power and societal geographies. Robust as these arguments might be, they leave unexplored both the nature and the corollaries of the relationship between the financial and the real production level.

3.2 The Institutional Perspective: Looking beyond the Real Production–Financial Profit Nexus

Turning to the institutional approaches to the study of financialisation, one can immediately identify the structural differences from the systemic ones: the nature of decoupling of real production from financial profit takes a back seat to the study of the embedded relationship between these two entities as well as the actors, agents and signs of financialisation. The main point of departure of this type of approach is that the term financialisation is conceived as expressing the engagement of non-financial market in the financial sector. In the context of the evolution of the capitalist system, this is viewed as the connecting thread that runs through the real economy, finance and society (Nesvetailova 2007).

The French Regulation School takes pride of place in this particular analytical tradition. Its contribution on the question of financialisation owes much to the works of Robert Cox (1987) and Susan Strange (1997), who grappled directly with the issue of the decoupling of finance from real production. Their arguments are based on the ideas of the institutionalisation of finance and social organisation. Strange suggested that the rise of financial instruments observed over the past decades had institutionalised the financial sphere to such an extent that it was able to disengage from the real economy and create profits without production. Cox, on the other hand, maintained that finance is central to understanding the status of the world order and that access to finance is an expression of collective social action as well as a statement of actors' material capabilities.

For the Regulation School, the accumulation regime is as important as the mode of regulation in the system. According to Elam (1994), one of the most notable writers of this school, the main advantage of the Regulation School is that it attempts to overcome the limitations of traditional mechanical approaches to the capitalist system. Instead of focusing on the idea of value, Regulationists shifted their analysis to the various social forms of capital and the influence of institutions. Aglietta and Lipietz, the creators of this theoretical tradition, claimed that the Regulation School has the merit of explaining "the paradox in capitalist development between the inherent tendency towards instability and crisis, and its ability to stabilise for periods around a set of institutions, norms and rules that secure periods of economic stability" (Webb 1998).

In this sense, the Regulation School proceeds from an analytical perspective that is more endogenous to the capitalist system, focusing on the institutions that have the power to influence it, hence, on why, rather than how, capitalism operates the way it does at any point in time. Their argument is based on the idea that regulation is a requirement for the stabilisation of an ever-volatile and fragile capitalist system; consequently, the social and political struggles of capitalist societies are responsible for the development of the necessary regulatory institutions. In addition, it claims that capitalism, just as society and politics, evolves over time and thus requires new institutions to develop in tandem in order to manage and sustain each phase of what they refer to as 'capital accumulation'.

It follows that each capitalist phase is uniquely identifiable by the type of regulatory institution and capital accumulation active at that particular point in time. Attempting a more sophisticated understanding of the term, Lipietz defines capital accumulation as "the parallel development over a long period of the conditions of production, such as the degree of mechanisation, and the productivity of labour", plus the social use of production through government expenditure and consumer spending (1992, p. 2).

All the above demonstrates that the Regulation School conceives financialisation as a dual process that involves both a shift in financial investment by non-financial institutions and a change in the overall macroeconomy. This approach views the growth of finance in a way comparable to Fordism, representing historical phases of expansion of the capitalist system, supported by their respective institutional forms as well as by accumulation and regulation regimes (Aglietta 1979). Unlike previous phases, however, the latest evolution is argued to result from the so-called 'dynamic' process of financialisation (Macleod 1997). In this regard, Aglietta points to the evolution of corporate governance and the finance regime to demonstrate, for example, the increasingly important role of banks in the era of financial liberalisation. Just as importantly, this approach shows that, while financial markets operate under the logic of homogenisation, the value of firms is not only dependent on share prices, but also highly leveraged, which creates a mismatch between productivity increases and aggregate demand.

From a more traditional institutional perspective, Froud et al. conceptualise the shift in corporate strategy and governance as a consequence of the rise of the shareholder value in the 1980s. In their words, "financialisation does connote important real changes, but is not immanent, economy wide principle and is not a coherent, realizable project for management" (2000, p. 104). This claim is in plain contradiction with the Regulation School, insofar as it precedes from the argument that financialisation, much as it may aid macroeconomic performance, cannot induce a convergence of corporate and economic growth. In this view, the emergence of a whole range of individual investors is no guarantee of institutional concurrency between the reforms shaped by corporate management and the reforms required by the capital markets.

At the macroeconomic level, most of the studies converge on the point that the process of financialisation can be defined as a finance-driven growth regime. Few of them, however, pay much attention to the institutional channels that capture the effects of this process, quite contradictory at times. On the one hand, financialisation is generally assumed to decelerate the levels of accumulation (Van Treeck and Hein 2007), but, at the same time, increased financial profitability is also often viewed as a driver of higher investment levels (Bhaduri and Marglin 1990). On the other hand, financialisation is understood to increase consumption levels due to real consumption, real or financial wealth, increased financial credit availability and increased dividend revenues (Cordonnier 2006). At the same time, however, the deceleration of real production is argued to lead to a reduction of total wages and, consequently, of consumption levels.

Another prominent institutional approach, made even more prominent in the wake of the rise of neoliberalism, is Post-Keynesianism. This approach centres the study of financialisation on the concept of rentiers, and especially the role of moneylenders in this capacity. Most authors associated with this school, including Crotty, Pollin and Epstein, seem to betray a broad Marxist slant in their claims about the negative impact of rentiers on the health of the real economy.

In particular, Crotty argues that the existence of the rentier inevitably reduces the level of profits available for investment and, therefore, the profits of industrial capitalist actors. From a Post-Keynesian perspective, the emergence of financialisation has created imbalances in the levels of real economic performance and investment that can only be redressed with some form of government intervention. This may come in the form of regulations on credit, financial investments, risky market involvements and liquidity levels in the system as well as in bank reserves (Crotty and Epstein 2009). In the Post-Keynesian view, this interactive kind of corrective government policy is seen as a necessary response to the central, and mostly pernicious, role that rentiers – be they speculators, price manipulators, or risk-inclined actors – play in the process of financialisation.

3.3 Could the Institutional and the Systemic Approaches be Bridged?

One could argue that a number of studies succeeded in offering a middle-ground perspective on financialisation, bringing together the two quite different accounts of it. An approach of this kind would suggest such definitions of financialisation as, for instance, a process whereby we are witnessing an "increasing importance of financial markets, financial motives, financial institutions, and financial elites in the operation of the economy and its governing institutions, both at the national and international level" (Epstein 2005). In this definition, Epstein incorporates not only the role of the institutional and systemic dimensions –both at the domestic and international levels – but also that of the evolving relationships within the society itself. Here, financialisation is conceived as a process that introduces new financial motives and actors and intensifies the role of existing financial markets, while transforming the institutions of the current regime to suit them. Epstein also adds that, as financial investments take over real production investments, the overall weight of financial activities in the system increases accordingly, in line with what Stockhammer (2004) identified as the growing primacy of shareholders' interest in firms.

Ben Fine (2009) takes a step further with a more inclusive Post-Keynesian conception of financialisation. In his reconstruction, the process of financialisation is the end result of six distinct, but inter-related, developments that unfolded over the course of the past four decades. These are (1) the growth and spread of financial markets; (2) the rapid development of speculative assets and products; (3) the simultaneous stagnation of real economic investment levels; (4) the development and increasing popularity of financial instruments and services; (5) the progressive loss of power of the industrial sector; and (6) the rising influence of finance. Additionally, Fine highlights the important role of financialisation as a process of debt-fuel consumption as well as of wealth redistribution and elite restructuring to the advantage of the international rentier class.

Having looked across the wide spectrum of accounts of financialisation, it becomes apparent that, compared to the historical and structural approach, this perspective places considerably more emphasis on the internal functioning and spillover effects of the relationship between real production and financial profit. In doing so, they produce valuable insights into the nature of this relationship and, in particular, draw long overdue attention to the underlying factors embedded therein, such as the ideas of regulation and shareholder value. This approach, therefore, focuses far more on the practical rather than the notional aspects of financialisation and, as such, neglects the more strictly social and structural dynamics of the process.

References

Aglietta, M. (1979). *A theory of capitalist regulation: The US experience*. London: New Left Books.

Altvater, E. (1997). Financial crises on the threshold of the 21st century. *The Socialist Register*.

Amin, S. (1980). *Class and nation: Historically and in the current crisis*. New York: Monthly Press Review.

Arrighi, G. (1994). *Il Lungo XX° Secolo: Denaro, Potere E Le Origini Del Nostro Tempo*. Milano: Il Saggiatore.

Baran, P., & Sweezy, P. (1966). *Monopoly capital: An essay on the American economic and social order*. New York: Monthly Review.

Bhaduri, A., & Marglin, S. (1990). Unemployment and the real wage: The economic basis for contesting political ideologies. *Cambridge Journal of Economics*, 14(4), 375–393.

Boyer, R. (2013). The global financial crisis in historical perspective: An economic analysis combining Minsky, Hayek, Fisher, Keynes and the regulation approach. *Accounting, Economics and Law – A Convivium*, 3(3), 93–133.

Carruthers, B. G. (2015). Financialization and the institutional foundations of the new capitalism. *Socio-Economic Review*, 13(2), 379–398.

Cordonnier, L. (2006). Le Profit Sans L'accumulation: La Recette Du Capitalisme Dominé Par La Finance. *Innovations Cahiers d'Economie De l'Innovation*, 23(1), 51–72.

Cox, R. (1987). *Production, power, and world order*. New York: Columbia University Press.

Crotty, J., & Epstein, G. (2009). Avoiding another meltdown. *Challenge*, 52(1), 5–26.

Elam, M. (1994). Puzzling out the post-Fordist debate: Technology, markets and institutions. In A. Ash (Ed.), *Post-Fordism: A reader*. Blackwell: Oxford.

Epstein, G. (2005). *Financialization and the world economy*. London: Edward Elgar.

Fine, B. (2009). *Financialisation and social policy. Social and political dimensions of the global crisis: Implications for developing countries*. Geneva: UNRISD.

Foster, J. (2008). The financialization of capitalism. *Monthly Review*, 59, 1–12.

Froud, J., Haslam, C., Johal, S., & Williams, K. (2000). Shareholder value and financialization: Consultancy promises, management moves. *Economy and Society*, 29(1), 80–110.

Hall, P., & Soskice, D. (2001). *Varieties of capitalism: The institutional foundations of comparative advantage*. Oxford: Oxford University Press.

Harvey, D. (1982). *The limits to capital*. Chicago: University of Chicago Press.

Hilferding, R. (1910). *Finance capital*. London: Routledge.

Jorda, O., Schularick, M., & Taylor, A. (2011). Financial crises, credit booms and external imbalances: 140 years of lessons. *IMF Economic Review*, 59(2), 340–378.

Krippner, G. (2005). The financialisation of the American economy. *Socio-Economic Review*, 3(2), 173–208.

Lapavitsas, C. (2009). Financialised capitalism: Crisis and financial expropriation. *Historical Materialism*, 17(2), 114–148.

Lapavitsas, C. (2010). Financialisation, or the search for profits in the sphere of circulation. *Economiaz*, 17(2), 114–148.

Lazonick, W., & O'Sullivan, M. (2000). Maximizing shareholder value: A new ideology for corporate governance. *Economy and Society*, 29(1), 13–35.

Lipietz, A. (1992). *Towards a new economic order: Post-fordism, ecology and democracy.* Oxford: Polity Press.

Macleod, G. (1997). Institutional thickness and industrial governance in Lowland Scotland. *Area*, 29(4), 299–311.

Magdoff, H., & Sweezy, P. (1969). *The deepening crisis of US capitalism.* New York: Monthly Review Press.

Nesvetailova, A. (2007). *Fragile finance, debt, speculation and crisis in the age of global credit.* Hampshire: Palgrave Macmillan.

Palley, T. (2007). Financialization: What it is and why it matters. *Levy Institute Working Paper*.

Parboni, R. (1980). *The dollar and its rivals: Recession, inflation and international finance.* London: Verso.

Paulani, L. (2009). The autonomization of the truly social forms in Marx's Theory: Comments on money in contemporaneous capitalism. *ANPEC Congress*. Rio De Janeiro: ANPEC.

Pike, A., & Pollard, J. (2010). Economic geographies of financialisation. *Economic Geography*, 86(1), 29–51.

Rappaport, A. (2005). The economics of short-term performance obsession. *Financial Analysts Journal*, 61(3), 65–79.

Rowthorn, B. (1980). *Capitalism, conflict and inflation.* London: Lawrence and Wishart.

Smart, A., & Lee, J. (2003). Financialization and the role of real estate in Hong Kong's regime of accumulation. *Economic Geography*, 79 (2), 153–171.

Stockhammer, E. (2004). Financialisation and the slowdown of accumulation. *Cambridge Journal of Economics*, 28(5), 719–741.

Strange, S. (1997). *Casino capitalism.* Manchester: Manchester University Press.

Van Treeck, T., & Hein, E. (2007). Dünhaupt P., Finanzsystem Und Wirtschaftliche Entwicklung: Neuere Tendenzen In Den USA Und In Deutschland. *IMK Studies*.

Webb, R. (1998). Regulation theory or Marxism: A consideration of two theoretical approaches to industrial relations in New Zealand. *University of Auckland*.

CHAPTER 4

Finance and the Oil Market: Introducing a Comprehensive Approach to Analysing the Financialisation Process

Abstract Having established that the concept of financialisation escapes narrow definitions and that financialisation, as a process, can only be accurately approached as a set of transformations within the global economic and financial systems, this chapter introduces a two-pillar approach which comprises analytical tools from both approaches to financialisation. This two-pillar approach developed in this book is designed to best capture the transformative impact of the process of financialisation, as it allows to gradually analyse the key elements of it: the behaviour and expectations of the financial actors, as well as the performance of both the physical and the financial dimensions of the oil market. Both phases of the analysis are conducted against the background of the changes in regulation and technology that conditioned the process of financialisation, owing to the particular institutional and socio-economic makeup of the oil market.

Keywords Financialisation · Economy · Financial economy · Real economy · Efficiency market theory

The previous chapter demonstrated that, not unlike the related notion of globalisation, the concept of financialisation indeed escapes narrow definitions, while any attempt to capture its processes and effects is bound to capsize on either the descriptive or the analytical side. One may conclude that financialisation, as a process, can only be accurately approached as a set of transformations, rather than a single and an impartible process (Casey 2011). Conceiving financialisation as a set of transformations within the global economic and financial systems will make it possible to capture the underlying dynamics of its evolution, both on the systemic and the institutional level.

With this in mind, the analysis offered to the reader in order to examine both the systemic and the institutional dimension of the process of financialisation in the context of the oil market, this book adopts a two-pillar approach. This approach comprises analytical tools from both of the approaches to financialisation. This combination is necessary due to the complex functioning of the financialisation process on the ground. The latter is largely determined by two forces: (1) the performative cycle between financial facades and underlying real economic markets and (2) the continuous evolution of technologies and regulations.

To that end, the institutional approach has the merit of showing the extent to which oil-based financial markets are embedded in the underlying oil commodity market. This link is contingent on the performance of the underlying market insofar as financial actors form their expectations, and therefore make their investment choices, on the basis of real economic performance indicators (Platts Energy Risk Forum 2012). At the same time, this link is reciprocal in nature, as the performance of oil-based financial products determines, in its turn, the performance of the underlying market. This circular cause-and-effect pattern is defined by the fact that physical markets view the performance of commodity derivatives[1] as the most reliable price forecast for their evaluations and investment decisions. The emergence of this performative cycle, as part of the process of financialisation, is a pivotal point of departure for the argument developed in this book.

Indeed, this development is responsible for the transformation of the structure of the underlying commodity markets, as it leads to the emergence of a new group of actors playing a crucial role in oil price determination. In order to trace the process of financialisation within the structure of a particular market, such as that of oil, it is necessary to first reconstruct its main determinants and distinct features, which will be done in the following chapter. Financial markets are known to blossom in the wake of deregulatory policies and other investment opportunities that encourage the participation of financial actors and the subsequent creation of financial products. This, in turn, predetermines the emergence of a new group of actors in the market, together with the changes in the commodity pricing mechanisms. The interests that drive this group of actors are purely financial, and are based on the performance of the market rather than the underlying commodity that is traded. This aspect is central to the process of financialisation. In addition, the flexibility of the new group of financial actors, in terms of their ability to enter or leave the market at will, translates into increased volatility in the market price level. Specifically, the price level inflates at times of positive expectations, as financial funds enter the market, and deflates at times of negative expectations, as financial capital flees the market. Not surprisingly, owing to the short-term – and, hence, often short-sighted – nature of most investment strategies, high volatility soon becomes a common, if not standard, feature of the market price level.

This volatility, however, can be analysed and understood through the financialisation lens, instead of being left as an independent variable in the functioning of the market. The argument of this book proceeds from the idea that the process of financialisation has paved the way for the participation, and growing influence, of financial actors and structures in markets where they were previously either absent or merely peripheral. The ensuing structural transformations can take the form of new institutions, new regulations, new commodity pricing mechanisms or even new parallel markets based on the real commodity. The actors responsible for this transformation are either pre-existing market participants, who choose to engage into, or increase, their financial operations to maximise profits or hedge risks, or actors previously unrelated to the market, who are attracted by its novel financial dimension. A further round of transformations occurs as these markets become increasingly vulnerable to the norms and institutions that the new actors bring along, on the basis of their behaviour and motives, on the original structure of the market.

The two-pillar approach developed in this book is designed to best capture the transformative impact of the process of financialisation, as it allows to gradually analyse the key elements of it: the behaviour and expectations of the financial actors, as well as the performance of both the physical and the financial dimensions of the oil market. Both phases of the analysis are conducted against the background of the changes in regulations and technologies that conditioned the process of financialisation, owing to the particular institutional and socio-economic makeup of the oil market.

This part of the book is devoted to setting out the relevant concepts, which will form the three elements of the methodological toolkit for deconstructing the process the oil markets' financialisation. First, the approaches to the relationship between the financial economy and the real economy will be examined, along with the concept of performativity and performative cycles. This will allow to further analyse how new actors enter financial markets and trace the reciprocal influence that the two exert on each other. The next step would be interpreting the behaviour of these new actors in the market – as well as their interactions with the 'traditional' market participants – through the lens of the main theoretical approaches to this issue. To complete the toolkit, a method for integrating technological and regulatory changes into the analysis of the financialisation process will be introduced at the final stage.

Ultimately, this methodology is designed to be instrumental for the analysis that will follow in Part II of the study, which will focus on the three phases of the financialisation of oil markets, cutting through the complexities of this process.

4.1 Analysing Financialisation: Financial Economy vs Real Economy, Conceptualising Performativity and the Emergence of New Financial Actors

As the process of financialisation entails the decoupling of the financial economy from the real economy, the study of financialisation must proceed from the study of the relationship between these two entities. Generally speaking, the link between stock markets and macroeconomic indicators has long been the object of extensive theoretical debates. This question is of particular interest to both the field of finance, in its quest to predict future stock market trends, and the field of macroeconomics, in its attempt to understand how the economy affects and is affected by the stock market.

Many studies have attempted to identify the link between stock prices and economic activity. Some, for instance, suggest that forecast price returns are strongly linked to the business cycles, as higher returns are expected in bad economic times and lower returns in good ones (Fama and French 1989). This study finds that bad economic times make households more risk averse, while good economic times are more risk tolerant, which in turn leads to higher inflation in good times as higher risk premiums lead to higher profits.

Other research endeavours (Estrella and Hardouvelis 1991) have discovered that a number of macroeconomic variables usually used to predict future stock returns could just as effectively be used to forecast the levels of economic activity. These variables include the ratio of investment to capital (Cochrane 1991), the ratio of dividend to earnings, investment plans rates, the ratio of labour income to total income (Menzly et al. 2004), the ratio of housing to total consumption (Piazzesi et al. 2004), the 'output gap' formed from the Federal Reserve capacity index (Cooper and Priestley 2005) and the ratio of consumption to wealth (Lettau and Ludvigson 2001). The most reliable indicators are considered to be the ratio of investment to capital and the ratio of consumption to wealth, as they are based on the idea that both firms and individuals will invest more when expected returns are low (Cochrane 1991).

With regards to the link between levels of investment and stock returns, Lamont (2000) found that investments react with a lag to changes in stock prices, as investment decisions take some time to get planned, approved and acted upon. Consequently, he opts for the investment plans indicator, and observes that their levels react almost instantly to changes in stock prices. Using investment theory to explain asset price anomalies, Zhang (2004), in contrast, finds that firms with low-expected returns to capital tend to invest more and to increase their exposure by selling more stocks, while firms with high-expected returns tend to repurchase their own stock. Merz and Yashiv (2005), on the other hand, study the possibility of a better fit of the investment to capital model when labour costs are included. The results of their study confirm that labour flow, and specifically the interaction between labour and investment, correlates well with stock market performance.

Barrell and Davies (2004) adopt another approach employing a vector-error correlation mechanism to study the links between real equity prices, real interest rates and government surplus levels as a percentage of Gross Domestic Product (GDP). Their findings show that in both the European countries they tested, as well as in the USA, there is a strong correlation between these variables. However, they conclude that the output levels of market-based economies are significantly more dependent on real equity prices, while real interest rates are universally negatively related to both output and equity prices. In a similar study, Pesaran et al. (2004) focused on the links between domestic equity prices and levels of GDP in a sample of twenty-five countries and found a strong, if variable, correlation.

In attempting to answer the question of whether and how the financial market is linked to the macroeconomic performance, these studies prove that certain macroeconomic indicators are indeed strongly affected by the behaviour of equity prices, especially inflation rates, interest rates, investment levels and output. The question that remains open, however, is what kind of links and mechanisms exist between macroeconomic and financial performance, and how they can be applied to the study of financialisation. Part of the answer to this question can be found in the theory of performativity.

4.1.1 Performativity and Performative Cycles

Coming from a more sociological perspective, the theory of performativity offers an analytical framework for the study of actor behaviour and can therefore be adopted into a political–economic perspective on the role of expectations and the performance of fundamental indicators.

The concept of performativity can be approached from two different theoretical fronts. From a strictly semantic perspective, performativity is related to the meaning of the word 'performance' and, as such, refers to the special quality of 'doing' – that is, the very act of making things happen, as opposed to contemplating, observing, declaring or representing them. In this view, language, discourse and any other type of expression are performative if and when they represent a real-life action that exists in its very performance. In the words of Judith Butler, the cultural theorist, performativity is the ability of something to be performative, that is, to create itself in the process of being expressed. In her Gender Trouble, for example, she uses the notion of gender to present a paradigmatic definition of the meaning of performativity, "There is no gender identity behind the expressions of gender [...] identity is performatively constituted by the very 'expressions' that are said to be its results" (Butler 1990, p. 25).

The second theoretical front defines the concept of performativity from an economic and sociological perspective. Performativity is used to conceptualise economics as the science that expresses the functions, behaviour and workings of the economic system and, therefore, as a discipline that performatively constitutes its own performance (Callon 1998). More specifically, Callon argued that economics is one of those practices that perform markets, thereby implying that the models and theories of economics find actual use in economic practice. Callon does not entertain the notion that economic theories can ever be completely disconnected from the workings of the economy itself; on the contrary, he contends that economists are actively engaged in the everyday practice of the economy, as they shift from the act of studying it to the act of performing it.

In disagreement with Callon, Daniel Miller (2002) argues that economists have not lost their theoretical vantage point but, rather, that the real economy has, itself, interjected the economists' theoretical models. In Miller's view, the concept of homo economicus, originally a pure figment of theoretical fiction, has gradually become reality on account of the fact that economic actors, once instilled with this particular belief, reconstituted the world around them in its image and, hence, performed it into reality.

> While the economic models [...] do not exist in reality, they are – increasingly – projected onto real-life economic behaviour with such force that people take them as objective, natural, thing like, and outside society, in much the same way as they think of gender, illness, death, and the laws of physics. (Holm 2007)

MacKenzie develops the economic approach to performativity even further. By drawing a distinction between the 'generic' and the 'effective' type of performativity, he contends that any theory, idea or model, in order to be effectively, rather than generically, performative, needs to go well beyond the observation of the economic process it was created to describe – it needs to be employed in a way that would actually affect and change it. He concludes that for economic theories to be performative, they must be introduced within the processes they were made to describe in such a way as to result in a variation of the functions and performance of these processes as compared to what they would have been in the absence of said theories. In his words, "To claim the economics is performative is to argue that it does things, rather than simply describing (with greater or lesser degrees of accuracy) an external reality that is not affected by economics" (Mackenzie 2007).

The concept of performativity can therefore be easily related to the concepts of actors' behaviour, expectations, perceptions and self-fulfilling prophecies. If economic performativity determines the way theories shape actors' understandings of the economy, and in turn, the way actors form their expectations on the basis of these understandings, then it is possible to argue that, by acting on these expectations, the actors end up shaping the economy in the image of these theories. As a result, these theories effectively become self-fulfilling – or, in the case of counter productivity, self-destructing – prophecies.

The example of a bank run is frequently used to that effect. MacKenzie, for instance, illustrates how, if rumours of liquidity problems at a bank spread through the economy, those who have their funds deposited in the bank will attempt to withdraw them because liquidity problems are expected to lead to bankruptcy and, eventually, to the possible loss of funds. However, this process deepens the liquidity problems of the bank and renders it unable to action all of the required withdrawals, thereby ensuring that it will, in fact, go into bankruptcy. This is a textbook example of an economic self-fulfilling prophecy, whereby economic theory forms the basis of expectations that are, in turn, acted out in a self-fulfilling loop.

By the same token, the link between the performance of stock markets and the macroeconomy depends on the expectations of both individual investors and firms with respect, this time, to the overall economy. This is because stocks, and the market for stocks, operate similarly to any other economic market. The price of stocks depends on the level of demand for stocks in the market, and it fluctuates along with it. The reason why demand for stocks is more volatile than in most other economic markets is that demand for products, and the willingness of people to pay the price to purchase them, depends on the level of expected utility.

In the case of stocks, however, expected utility coincides with expected profits, which in turn, as discussed above, depend on a very delicate combination of heuristics, cognitive evaluations and subjective feelings. As a result, if the expectation of a downturn in the stock market is dismissed or overlooked by the public, then the effect of a negative change in macroeconomic performance will be smaller than if panic and loss of confidence had prevailed. In contrast, if euphoria prevails, then the expected positive effects of an upturn in the macroeconomic performance will be inflated. Although only one side of this direct relationship between stock market and macroeconomic performance is commonly acknowledged, the direction of dependency goes both ways.

Indeed, it is not as often acknowledged that individuals and firms, who both exist and operate within what is considered to be the macroeconomy, are also directly affected by the performance of the stock market.

Individuals, when involved in financial markets, usually allocate a part of their savings to an investment portfolio that they expect to yield certain returns over time. Individual investors build and treat their portfolios in different ways and with a different tolerance to risk and diversifications. What is common to most of them, though, is their reaction to the stock market performing well: individuals are confident, hence, they are spending more, saving less and investing either in the stock market itself or, guided by the general expectation of high returns, in other unrelated markets (Fama & French, Business Conditions And Expected Returns On Stocks And Bonds 1989). This behaviour is very closely linked to the concepts of wealth effect and money illusion (Akerlof and Shiller 2009), as first described by Minsky and Keynes, which suggest that the performance of individuals' portfolios has a direct influence on their approach to the market because, when their investments increase in value, their confidence in their own wealth and expenses increases accordingly – and vice versa. The money illusion also refers to a psychological effect that is closely linked to the economic education of investors; that is, the failure to factor in the effects of inflation in their investment decisions.

As a result, positive performance leads to increasing portfolio values and, consequently, to increasing confidence, increasing spending and increasing investment activity. This behaviour has inflationary consequences on the macroeconomic indicators, and usually requires corrective, viz. reflationary, monetary policy. Similarly, if the stock market is performing poorly and stock prices are in free fall, individuals tend to feel worse off; they lose confidence and, as a result, tend to spend less and save more. Increased savings and reduced consumption create a negative environment for investment, which, again, has a noticeable impact on the macroeconomic equilibrium.

In contrast to the case of individual investors, the way firms are affected by the stock market is complicated by the fact that, in any advanced economy, most large firms, not only are themselves part of the stock market, but also trade in their own stocks. This creates two routes through which the stock market can affect the macroeconomic equilibrium. First, most firms, just like individuals, invest parts of their capital in stock markets and other financial market portfolios. This means that if the stock market is booming, their assets rise in value and firms feel confident to expand either against hostile takeovers or towards ambitious business ventures. In doing so, the growth of firms will translate into increased productivity not just for the firms themselves, but also for the whole economy. Conversely, a poor stock market performance will diminish the value of firms' assets and make them less likely to expand and, hence, to generate macroeconomic growth. This mechanism is even more conspicuous in the case of institutions, such as banks, pension funds and investment organisations, whose operations depend to a great extent on the performance of the stock markets as a large part of their capital goes into investments.

Another way in which financial performance of firms reflects the relationship between stock markets and macroeconomic performance is through their own stocks. When the stock price of a firm falls, so does the confidence of its shareholders. Lower prices on the stock of a firm mean that the value of the firm is shrinking – usually because of unforeseeable events, the introduction of a competitor, mismanagement or just plain speculation. When the value of a stock falls, its holders will usually sell it immediately in the expectation that the prospects of the company are going to worsen. If the firm, then, wishes to increase its liquidity in order to maintain the same level of funds, it will be forced to release larger stakes of its shares. This loss of value and confidence can make firms vulnerable to hostile takeovers and, in some cases, even liquidation. Although individual cases have a negligible impact on the overall macroeconomic performance, this condition, especially when the stock market is in a particularly vulnerable state and many stock prices are falling in value, can spread to whole sectors. In that event, the effects of the contagion are felt throughout the whole economy as there is no good performer to cancel out the loss in value and the reduction in investment activity.

Similar dynamics have been observed in the performativity literature. The performative cycle shapes actors' perception and understanding of the economy and the markets, and allows them to form their expectations and behaviour within its structure, while influencing their performance through their investment decisions. Therefore, while the case for the decoupling of the financial economy from the real economy still stands, these two entities must be recognised as being heavily interdependent. This has to do with the fact that financial actors form their expectations and investment decisions on the basis of real economic indicators, while at the same time, real economic indicators are heavily influenced by the financial markets.

Insofar as the performative cycle answers the question on the links and mechanisms underpinning the relationship between the macroeconomic and financial performance in the context of financialisation, both the financial markets and the real economy can be scaled down to the narrower analytical space of the 'financialised' market and the market performance of the underlying commodity product. More specifically, the analysis conducted so far indicates that the kind of financial actors and structures that emerge from the process of financialisation gives rise to an entirely new market dynamic. Here, the market is determined more by the norms and behaviour of the new financial actors and structures than by the laws of demand and supply that typically regulate non-financial markets, and hence become 'financialised'.

In financialised markets, the dominance of the new financial actors has to do with their ability to shape the performance of the underlying market and while being reciprocally influenced by this process itself. This way, the new financial actors become themselves part of the performative cycle between their own behaviour and the price determination dynamics of the underlying market. It is suggested that the emergence of a fourth 'financial' actor in the oil market is linked to just this type of performative cycle, particularly in view of its increasing influence in the price-setting mechanism of the oil market. As the distinctive feature of a 'financialised' market, this explanation could account for the historical process of financialisation that has left the oil market increasingly exposed and connected to international financial markets.

4.2 Analysing Financialisation: Behaviour of the Financial Actors

If the new market dynamic is determined by the expectations and behaviour of the new financial actors, it remains to be seen what this behaviour entails in practice. This question is central to understanding how markets change as the process of financialisation takes root within their internal structure. The differences between the behaviour of new financial actors and that of the traditional economic actors' active in the underlying commodity market may help bring to light the distinctive evolutionary features of a market under the process of financialisation. The study of financial actors can be approached from two main theoretical perspectives: the neo-liberal Efficiency Market Theory (EMT) and the Post-Keynesian theory.

EMT boasts a rich academic tradition. Among the most notable contributors to the EMT literature are Eugene Fama (1970) and Paul

Krugman (1983), who have approached the study of financial crises with the vocabulary of classical economics, which notoriously treats financial markets as self-regulating and inherently efficient. There are two generations of EMT models, which reflect the various revisions and developments that have been made over time to adapt its main argument to the evolution of the financial markets.

Based on the neo-classical economic tradition, this theory assumes that financial markets are no different than any other market; that is, they are inherently efficient. Paul Samuelson's (Samuelson 1963) classic take on the efficiency of the goods market is that individuals tend to spend money in order to purchase what they wish. If demand for some goods increases, their price will rise accordingly and so will production, until material resources run short and cause prices to rise even higher. On the other hand, if the availability of some goods exceeds the quantity demanded, prices will drop as suppliers hasten to sell as much of the goods as possible. As prices drop, the appeal of the goods will increase, and so will demand, until demand meets supply. In this theory, market prices need to be very volatile for the price-setting mechanism to set in and keep demand and supply in equilibrium. Not surprisingly, according to the EMT, asset prices are always at their optimal level, as they always reflect the true value of their investment. Last but not least, EMT, on the assumption that the financial markets bear no difference to the goods markets, also seamlessly incorporates the ideology of laissez-faire economics.

Uncritical belief in laissez-faire is one of the major flaws most frequently attributed to EMT, which is especially criticised for not entertaining the existence of irrational behaviour and for assuming that markets always operate in condition of equilibrium (Cooper 2008). These criticisms form the basis of the second theoretical approach to the issue of financial actor behaviour, known as the Post-Keynesian tradition. Kept for decades in the shadow of the Chicago School of Economics by both the academic and the financial world, this theoretical front has made a remarkable comeback in the past fifteen years. One of the merits of this theory is that it factors psychological and sociological insights into the study of economics and finance. In doing so, Post-Keynesians expose many of the long-established formulas and theories associated with classical economics as false or out-dated; most controversially, they conclude that people behave irrationally – and that markets do so, too (Sheleifer 2000).

Keynes argued that in financial markets the savings–investment nexus does not operate the same way as it does in other markets: whereas in typical markets savings affect investment levels, in financial markets it is investments that determine saving activity (Nesvetailova 2007). This is based on the assumption that when stock markets perform well, people will have an incentive to invest more than they would otherwise do. In apparent logical contradiction, however, this seems to imply that savings still do have an underlying influence over investment levels. Such apparent incongruities are not foreign to the explanatory framework of this theoretical tradition. The inter-relatedness among investment, savings and borrowing, in particular, is at the heart of Keynes' notorious paradox of thrift.

The ideas put forward by Minsky in the late 1970s and early 1980s became, in time, part of a certain Post-Keynesian canon (Fazzari 1989) – even though Minsky never considered himself as belonging to this theoretical school. One of his main arguments centred on the idea that, just as increased savings set in motion the whole paradox of thrift, increased borrowing produces a similar phenomenon, but with an opposite effect. In other words, higher borrowing is argued to increase investment levels, hence profits, and leads to a vicious circle where individuals save less, but borrow and spend more. As captured in perhaps one of his most eloquent quips, 'stability creates instability', he believed that at good times people tend to build up unsustainable amounts of debt, and that this tends to jeopardise the stability, or the otherwise desirable status quo, of the economy.

This argument has some echoes of Veblen's concept of 'conspicuous consumption'. Introduced almost a century earlier, this referred to the idea that people always seek to better their position in their effort to 'keep up with the Joneses' (Veblen 1899). This activity is indeed as conspicuous and it is consumptive because progress is both ostentatious and quickly forgotten, as people, in seeking to improve their apparent status, overstretch their purchasing power and inflate the economy.

Another important Post-Keynesian contribution is Mandelbrot's demonstration that markets have memory and that their performance is influenced by their own recent behaviour (Cooper 2008). Memory, he suggested, allows markets to readjust to their expected levels. This means that even though market conditions may change, actors contribute to their readjustment as they base their positions on their expectations as well as on the current condition of the market. This argument is clearly incompatible with EMT as it implies that market equilibrium is not determined as much by its true value, as by expectations. However, within the Post-Keynesian school, Mandelbrot's claims are pivotal. This is especially so because the market–memory argument seems to confirm the positive feedback process first elaborated by Minsky, which similarly posits that markets tend to repeat past events in cycles as if they had a memory of their own.

The attributes of actor behaviour identified in the Post-Keynesian theoretical framework throw much light on why, at times of increased risky investments with high returns, the booming financial market becomes so appealing to economic actors in spite of its high volatility. The main reason for the limited memory span observed in investor activity, according to Gorge Cooper, is to do with the main variables that investors employ to calculate the value of assets, to wit economic fundamentals, as discussed earlier in the chapter, balance sheets and income statements. All three of these variables are heavily linked to the performance of the financial market and, hence, their bias misleads investors into a false sense of control.

4.2.1 The Behavioural Finance Approach: From the False Sense of Control to 'Mob Psychology'

This phenomenon of a false sense of control is well known in the field of behavioural finance, and was most recently explored in Robert Shiller's and Akerlof and Shiller's Animal Spirits (2009). Behavioural finance offers a number of arguments relative to the formation of financial actors' expectations and behaviour. Their most cited arguments are that investors often trade on 'noise', rather than fully understood and pondered information (Sheleifer 2000); that they tend to rely on advice from both reliable and unreliable sources; that they fail to diversify their portfolios; and that they often sell well-performing stocks while holding on to underperforming ones (Sheleifer 2000, p. 10). These arguments are further explored and all appear to show that financial actors are seldom driven by rational calculations and incentives.

In his *Psychological Analysis of Economic Behaviour*, Katona puts in no uncertain terms that, in order for the rationality argument to hold any water, human beings must be assumed to be 'automatons' devoid of any human perception or feeling. In his view, this form of 'mechanistic psychology' has induced economic analysts to look at money and price behaviour strictly in terms of money and price, and to overlook the crucial human element therein. However, he argues that economic behaviour is driven by impulses all too human, for "how we spend our money depends on fashion, salesmanship and advertising, social background and standards, considerations of prestige, insecurities and emotional conflicts" (1951, p. 63).

In a similar vein, Loewenstein, in his *Exotic Preferences* (2008), investigates the process of investment decision-making and concludes that, because risks are conceived by investors as feelings, their outcome has a considerable influence on the future behaviour of financial actors. From a similar psychological angle, Paul Samuelson, in his *Risk and Uncertainty: A Fallacy of Large Numbers* (1963), introduces the concept of 'loss aversion' to refer to the phenomenon whereby actors behave myopically and irrationally in an attempt to minimise the risk of suffering losses. This behaviour has been linked to actors' attempts to build elaborate formulas and methods of beating the market. These forms of 'heuristics' or 'mental calculations', which usually involve the interpretation of signs, patterns and trends, are widely used among investors in the belief that they can help to predict the future behaviour of the market (Sheleifer 2000). Heuristic techniques of this sort instil investors with irrational overconfidence and usually lead to 'noise' trading (Shefrin 2007).

Noise trading can also be explained in terms of what Robert Shiller and Gorge Akerlof call 'animal spirits' (2009). In their study of actor behaviour, they posit that people's expectations are formed by the stories they hear. In other words, as people get involved in the markets in one way or another, they take into consideration what other people experienced in similar circumstances. Shiller and Akerlof go on to contend that investment decision-making is not only determined by purely economic or rational factors, but also by such animalistic spirits as herding, overexcitement, fear and panic. Two eminent hedge fund managers, George Soros (2008) and Peter Lynch and Rothchild (1990), have described the irrational determinants of investment decision-making in similar terms. Soros, in particular, maintains that financial markets are influenced by the 'reflexivity' factor, that is, the idea the actors who influence the financial markets through their activities are themselves influenced by the very markets of which they are a part, in a self-perpetuating loop that not only proves but also intensifies the link between human behaviour and financial market performance.

In *Manias, Panics and Crashes*, Charles Kindleberger proposes a six-level scale to measure the degree of rationality in the behaviour of the financial actors (2005). On level 1, actors are assumed to behave completely irrationally and to engage in 'mob psychology', 'groupthink' and other herding behaviour. On level 2, actors are expected to behave completely rationally, but only until they inevitably and progressively slide back to the first level. On level 3, different groups of actors conceive rationality in different ways; some of them behave rationally while others do not. On level 4, all the market participants renounce the 'fallacy of composition' and individual actors will conceive of rationality independently from the total of the group or the rest of the market actors. On level 5, the market operates under rational principles, but expectations differ on the quality and the quantity of information. Finally, level 6 is reserved to those financial actors who behave irrationally by either overlooking or suppressing key information as a result of misleading predicting models.

The use of this scale helps Kindleberger to show the prevalence of mob psychology and hysteria as an occasional, but well-established, deviation from rational behaviour and, to borrow Minsky's conception of 'optimism' and 'euphoria', that optimism can very easily turn into mania. The merit of Kindleberger's analysis is that it shows not just the prevalence, but also the dangers of the herding effect in financial markets.

The Post-Keynesian literature on the link between financial market performance and actor behaviour is virtually unanimous. The behaviour of financial actors is described as irrational, because they are influenced by 'animal' instincts, fear of losses, overconfidence, misleading heuristics, as well as peer pressure, herding and other marks of mob psychology. Financial actors are found to be myopic in their investment strategies and base their decision-making on expectations shaped as much by their own feelings as by the performance of the underlying economy.

Understanding the role of actors' expectations and irrationality is essential to the study of the financialisation of non-financial markets and the structural transformations that this process entails. The fact that financial actors' decision-making is so heavily swayed by such factors as herd behaviour, euphoria, panic or other market instincts is indicative that these actors may be less involved in understanding the underlying commodity of the market than in gratifying their own heuristic performance. It is the unpredictability and volatility of this behaviour that sets financial actors apart from traditional market actors.

In particular, even though financial actors do take into account the fundamental performance of the underlying commodity when forming their market expectations, they are usually more short-sighted than traditional investors. Not being tied to any aspect of the market itself, save for its performance indicators, financial actors tend to be biased in favour of short-term gains because they are free to enter and leave the market to the best of their interests. This is in stark contrast to the behaviour of traditional actors who, often motivated by a genuine interest in the underlying commodity or in its profit returns, make long-term commitments and investments in the market, and do not abandon it as easily. The more elastic – and volatile – behaviour of financial actors, in contrast, allows them to enter the market when they expect it to perform well and to leave it when it shows signs of sluggishness.

This form of short-term performance heuristics makes the financial performance of the market more volatile when coupled with instances of 'mob psychology' and 'herding' (Clapp and Helleiren 2010). Volatility is also further intensified by the ease with which funds are able to enter and leave the market, which, depending on the popularity of the market in question, end up inflating its price level. In turn, these pricing pressures are transferred from the commodity-based financial market to the underlying commodity market by the performative cycle that links them together. On top of this, the underlying commodity market is also more responsive to those expectations that performatively shape events into reality. These events, which include financial and economic crises, as well as instances of general euphoria, can increase the volatility of the behaviour of financial actors even when the underlying market is unaffected.

4.3 Analysing Financialisation: The Role of Technology and Regulatory Evolution

Having gone through the relevant economic fundamentals, the roots of the emergence of new financial actors on the market scene, as well as the approaches to their behaviour and expectations, the previous parts of the chapter offered a number of relevant methods and tools to analyse the relationship between real economic production and financial profit creation and demonstrated how it can be linked to the process of financialisation. This part of the chapter will add the final methodological element to the toolkit, integrating the factors of technology and regulation into the study. It will examine the extent to which these factors underpin the process of financialisation within and without the structure of a traditional commodity market. These two factors are dual in scope; they can be either universal or market-specific, according to the number of markets targeted by the technological or regulatory developments.

While such technological advancement is equal at the international level, when it comes to the market-specific one, the rate at which new technologies are adopted into the structure of a given market is not always in line with the rate at which they become available to the public or to other markets. Similarly, regulation exists at multiple levels, including the international, the national and the market-specific one, and it has the potential to influence extensively market performance, structure and functions. The process of financialisation exposes markets to advancements in financial technologies and regulation, even if these are not market-specific.

At the universal level, technological advances have profoundly affected the international financial sector.

> During the May 2010 flash crash, some stocks traded as low as a penny before recovering in a manic 20-minute period [...] It may not take a full trading day for the markets to lose 25 per cent today – it could happen in moments. And while traders knew trouble was brewing when they arrived for work on October 19, 1987, today companies can lose hundreds of millions of dollars out of nowhere. (Mackenzie et al. 2012)

The introduction of radio, television and finally the Internet has triggered sweeping changes in the world of finance as instant access to information has progressively broken down the physical barriers between financial operations (Guttmann 2002). Advances in communication and information technology have also changed the structure of financial markets by giving new actors the information tools to gain market access (Kurtzman 1993). With greater access to information, actors gained a new level of awareness and involvement in the market. Virtual data, in particular, has made it possible for actors to operate outside the physical and theoretical bounds of each market.

The advancement of new technologies, and their progressive adoption by the financial sector, has led to the creation of new financial products and markets. Among the latest technological breakthroughs that have made an impact on the financial markets is the Internet. Its introduction led to the electronic marketisation of financial products, with real-time access to information at an international scale. This development altered the dynamics of the financial markets in substantial ways, as international funds now had the ability to collect the necessary information to gain access to these markets, which suddenly received vast new volumes of investments. What is more, these new investments turned out to be more flexible than the traditional ones; their volumes could change very rapidly and this could intensify the reaction levels of the markets in both speed and magnitude.

However, these developments are not just the result of technological advances in financial markets, and of the ease with which actors can now access and exit them. Indeed, the literature on the behaviour of financial actors, discussed earlier in the chapter, shows the pervasive role of the human element in financial decision-making; technological advances have only intensified such patterns of behaviour as herding or mob psychology. Indeed, having access to the same information has made the possibility of experiencing the same reactions all the more possible and frequent.

This information revolution is all the more relevant when considering how technological advances can affect the structure of a non-financial market in the process of financialisation. While both financial and non-financial markets are equally affected by the advancement of technology, the effects of the latter on these two types of markets differ. Non-financial markets will stand to benefit from more efficient ways of production, better or faster supply and transportation routes, increased information, and faster communications; their profitability is likely to increase as a result. In contrast, in the case of the financial markets, increased profitability is not the only outcome of technological developments; flexibility is, too.

Financial actors value the role of information and communication in their business as much as they value the freedom to abandon a deflating market for a booming one. This form of flexibility is facilitated by the development of technologies such as the Internet, because this makes it possible to create, and access, new products without physical or geographic barriers. Therefore, technological advances, in the context of financialisation, not only lead to increasing profitability, but also to increasing volumes of capital investments and to the introduction of volatility risks directly determined by actor behaviour and indirectly allowed by technological accessibility.

Americans' battle with these complex (financial) trades is being closely monitored by regulators in Canada, Australia and the European Union (EU) as they too seek ways to contain volatility caused by machines. What frightens investors most of all is a sudden evaporation of liquidity, when everyone pulls back at once and there is no one to give a firm price to an investor wanting to sell (Mackenzie et al. 2012).

In this sense, the level of technology integrated into the structure of a financialised market determines the effects that the process of financialisation will have on the structure and functions of the market; it will do so by defining the level of engagement and influence of the new financial actors as well as by facilitating the creation and accessibility of financial products and markets.

Regulation, too, has the ability to condition the creation and accessibility of financial products and markets. However, in contrast to technology, regulation is reactive in scope and, as such, usually acts as a restriction to the otherwise available functions and accessibility levels. In general terms, the role of regulation is to target known problems and risks in financial or economic activities. Regulation can act at the international, national or market-specific level, and it can target either specific actors or entire markets. The process of financialisation can be affected by regulatory developments at any of these levels as long as the financialised market in question imports or deregulates activities within its structure.

Actor-targeting regulation refers to regulations imposed on the behaviour or activities of a specific group of actors in an attempt to minimise the risks to which they are believed to be exposed. Such regulation can take the form of minimum required liquidity reserves or even restrictions on high-risk investments. On the other hand, market-targeting regulation is designed to either prevent the creation of specific financial products or to regulate their size. Regulatory developments on either front have, therefore, the potential to affect both the creation and the accessibility of financial markets.

The deregulation of actors and markets facilitates the creation of new financial products and increases the accessibility of existing financial markets. In the case of financialised markets, regulation has a similar effect. Here, regulatory developments can pave the way for new financial products to be introduced into the financial structure and – unprecedented in these newly financialised markets – for capital to engage in them. In the absence of regulation, the financialisation process of a non-financial market would grow without any restrictions. Nevertheless, in the presence of regulation, any expansion causes instant and extensive movements of financial capital in the market that cannot be matched, at least in the short term, by the performance of the underlying market.

The role of regulation and technology in changing the structure and volume of the financial markets has also an impact on the geographical reach of the financialisation effects. The creation of new markets and the deeper involvement of new financial actors within the structure of financialised markets (Schornick 2009) tend to internationalise the effects of financialisation to other markets as these actors do not usually operate solely within the structure of one market, but across a multitude of markets to better hedge their risks. This creates links among the markets where these actors are most active.

Traditional non-financial markets consist of producers, consumers and intermediary firms. The economic performance of these actors is determined by the performance of the market, as they are confined to operate within this market on account of their capacities, investments or agreements. While the performance of their market affects them directly, any effects that their performance might have on other markets will only be indirect. In contrast, in markets undergoing financialisation, these dynamics change. The financial actors that enter the structure of the market are not as restricted by financial, physical or geographic barriers as traditional actors are. This allows them to partake in financial activities in different markets worldwide. As a result, the performance of a single specific market will affect their financial performance directly; however, it will also indirectly affect the performance of the other international markets and economies in which these actors are based or involved (Kyle and Xiong 2001).

With the introduction of technological advances and regulatory developments, the number of international actors with potential access to the structure of a market undergoing financialisation rises. This rise results in increasing global interconnectivity between this specific market and international financial markets. The link cuts both ways, as both the market undergoing financialisation and international financial markets are subject to the performance and preferences of the financial actors involved. The introduction of new technologies and regulations will bring about either a convergence or a disconnection of the two, depending on the nature of the underlying commodity of the financialised market. This is because some markets, such as the oil market, can be viewed as risk-hedging markets, insofar as the volumes of investments in their structure increase during financial downturns.

Hence, understanding the role of technology and regulation in the determination of the extent and effects of the process of financialisation in a market is crucial. It also adds important insights into the direct and indirect reach of the influence of financial actors, and of the relationship between real production and financial profits, as well as on the performance, structure and dynamics of a traditional non-financial market undergoing financialisation.

The integration of the changing dynamics of technological and regulatory evolution, therefore, completes the analytical toolkit that is required to proceed with the analysis of the three phases of oil market financialisation.

NOTE

1. Commodity derivatives date back to 1848 when a new type of investment was introduced in the Chicago Board of Trade by which insurance was provided to farmers as a means for protecting their farms through a promise to sell crops in the future for a pre-arranged price. This, however, created an investment tool that allowed investors to profit from certain items without possessing them, which has resulted to modern commodity derivatives trading to be most popular with people outside of the commodities industry.

REFERENCES

Akerlof, G., & Shiller, R. (2009). *Animal spirits.* New York: Princeton University Press.
Barrell, R., & Davis, P. (2004). Equity prices and the real economy – A vector-error-correction approach. *NIESR.*
Butler, J. (1990). *Gender trouble: Feminism and the subversion of identity.* London: Routledge.
Callon, M. (1998). *The laws of the markets.* Oxford: Blackwell.
Casey T. (2011). Financialization and the future of the neo-liberal growth model. Political Studies Association Conference Proceedings, 18–21 April, Terre-Haute, IN.
Clapp, J., & Helleiren, E. (2010). Troubled futures? The global food crisis and the politics of agricultural derivatives regulation. *Review of International Political Economy,* 19(2), 181–207.
Cochrane, J. (1991). Production-based asset pricing and the link between stock returns and economic fluctuations. *Journal of Finance,* 46(1), 207–234.
Cooper, G. (2008). *The origin of financial crisis, Central Banks, credit bubbles and the efficient market fallacy.* London: Harriman House Ltd.
Cooper, I., & Priestley, R. (2005). Stock return predictability in a production economy. *Norwegian School of Management.*
Estrella, A., & Hardouvelis, G. (1991). The term structure as a predictor of real economic activity. *Journal of Finance,* 46(2), 555–576.
Fama, E. (1970). Efficient capital markets: A review of theory and empirical work. *Journal of Finance,* 25(2), 383–417.
Fama, E., & French, K. (1989). Business conditions and expected returns on stocks and bonds. *Journal of Financial Economics,* 25(1), 23–49.

Fazzari, S. (1989). Keynesian theories of investment: Neo-post and new. *Revista Di Economica Politica*, 9(4), 103.

Guttmann, R. (2002). *Cybercash: The coming era of electronic money*. London: Palgrave Macmillan.

Holm, P. (2007). Which way is up on Callon? In Mackenzie, D., Muniesa, F., & Siu, L. (Eds.), *Do economics make markets? On the performativity of economics*. Princeton: Princeton University Press.

Katona, G. (1951). *Psychological analysis of economic behaviour*. London: McGraw-Hill Inc.

Kindleberger, C., & Aliber, R. (2005). *Manias, panics and crashes. A history of financial crises* (5th Edn). Hampshire: Palgrave Macmillan.

Krugman, P. (1983). Oil shocks and exchange rate dynamics. In J. Frankel (Ed.), *Exchange rates and international macroeconomics*. Chicago: University of Chicago Press.

Kurtzman, J. (1993). *The death of money*. London: Little, Brown and Company.

Kyle, A., & Xiong, W. (2001). Contagion as a wealth effect. *Journal of Finance*, 56(4), 1401–1440.

Lamont, O. (2000). Investment plans and stock returns. *Journal of Finance*, 55, 2719–2745.

Lettau, M., & Ludvigson, S. (2001). Consumption, aggregate wealth, and expected stock returns. *Journal of Finance*, 56(3), 815–849.

Loewenstein, G. (2008). *Exotic preferences, behavioural economics and human motivation*. Oxford: Oxford University Press.

Lynch, P., & Rothchild, J. (1990). *One up on Wall Street: How to use what you already know to make money in the market*. London: Penguin Books.

Mackenzie, D. (2007). Is economics performative? Option theory and the construction of derivatives markets. In D. Mackenzie, F. Muniesa, & L. Siu, *Do economics make markets? On the performativity of economics*. London: Princeton University Press.

Mackenzie, M., Massoudi, A., & Foley, S. (2012). Rage against the machine. *Financial Times*, October 18.

Menzly, L., Santos, T., & Veronesi, P. (2004). Understanding predictability. *Journal of Political Economy*, 112(1), 1–47.

Merz, M., & Yashiv, E. (2005). Labour and the market value of the firm. *University of Bonn*.

Miller, D. (2002). Turning Callon the right way up. *Economy and Society*, 31(2), 218–233.

Nesvetailova, A. (2007). *Fragile finance, debt, speculation and crisis in the age of global credit*. Hampshire: Palgrave Macmillan.

Pesaran, H., Scheurmann, T., & Weiner, S. (2004). Modelling regional interdependencies using a global error-correcting macroeconometric model. *Journal of Business and Economic Statistics*, 22(2), 129–162.

Piazzesi, M., Schneider, M., & Tuzel, S. (2004). Housing, consumption, and asset pricing. *University of Chicago*.

Platts Energy Risk Forum. (2012). London: Platts Media.

Samuelson, P. (1963). Risk and uncertainty: A fallacy of large numbers. *Scientia*, 69(1), 108–113.

Schornick, A. (2009). International capital constraints and stock market dynamics. *INSEAD*.

Shefrin, H. (2007). *Beyond greed and fear, understanding behavioural finance and the psychology of investing*. Oxford: Oxford University Press.

Sheleifer, A. (2000). *Inefficient markets, an introduction to behavioural finance*. Oxford: Oxford University Press.

Soros, G. (2008). *The new paradigm for financial markets, The credit crisis of 2008 and what it means*. New York: Public Affairs.

Veblen, T. (1899). *The theory of the leisure class*. Montana: Kessinger Publishing.

Zhang, L. (2004). Anomalies. *University of Rochester*.

PART II

The Three Phases of Oil Financialisation

CHAPTER 5

Oil Products and Oil-Based Financial Products

Abstract Having established that financialisation is a process of change in the nature and motives of market investment, market actors and market operations, this chapter embarks on an analysis of the emergence and evolution of the different types and functions of oil-based financial investment products. It then moves into a study of the new financial actors who have entered the oil-based financial market, by examining their different types, motives and practices while paying special regard to the technological and regulatory developments that have made this new market environment possible. This chapter also unveils the emergence of the process of financialisation of the oil market in terms of its progressive integration with the financial sector in the post-1990s period, through the analysis of three distinct phases of this process: the low (1990–1990), the early (1991–2001) and the advanced (2002–2015) financialisation.

Keywords Oil products · Oil-based products · Investment products · Oil market

As discussed in the previous chapter, the definition of financialisation employed in this book is that of a process of change in the nature and motives of market investment, market actors and market operations. In what follows, the analysis will be focused on the emergence and evolution of the different types and functions of oil-based financial investment products as well as of the different types, motives and practices of the new financial actors responsible for such capital investments, with special regard to the technological and regulatory developments that have made the new market environment possible.

This chapter will unveil the emergence of the process of financialisation of the oil market in terms of its progressive integration with the financial sector in the post-1990s period, through the analysis of three distinct phases of this process: the low (1990–1990), the early (1991–2001) and the advanced (2002–2015) financialisation.

The point of departure of this analysis should be however the year 1980, when a group of energy and futures companies founded the International Petroleum Exchange (IPE) which went on to release the first contract for oil futures in 1981. As Mabro (2005) observes, it would then take another five years for the first oil contract to employ a formula containing a spot price as a benchmark for the calculation of its price level. Even though oil had long been traded as a primary energy source, it was only after the events of the 1970s and the first oil crisis, in particular, that financial institutions realised its potential as an investment opportunity. In the aftermath of the 1973 oil crisis, the oil industry explored new methods for the diversification of risk and new capital. The methods associated with foreign exchange and agricultural products not only proved to be a viable solution, but also had the advantage of promoting transparency and attracting greater liquidity.

The development of the futures market provided openly available information on the current and expected conditions of the market to all its actors, making its functions more transparent. It also equipped market participants with the knowledge and ability to hedge, or shed risk, against unexpected price fluctuations. This proved especially beneficial to companies involved in the exploration, extraction, refining and even marketing of oil products. In case of a drop in oil prices, producers now had the ability to counterbalance their losses in the actual oil market with the profits gained by betting on said drop. To be sure, in the opposite case, their winnings from the actual market are offset by the losses in the financial investment. However, the introduction of the futures market made it possible for producers to reduce their vulnerability to the fluctuations of the oil price level through the use of oil future positioning.

Since the launch of the first crude futures contracts in the late 1980s (IEA 2011), the market has experienced all but continuous year-on-year growth. New products and trading tools have arisen in the meantime, notably swaps and options, and since 1999 trade of oil commodity products became available electronically on the Internet. Before proceeding with the analysis of the three phases of the financialisation process, it is worth providing a brief overview of both oil products and oil-based financial products relevant for this study.

As far as the commodities themselves are concerned, the energy sector trades in four distinct kinds of oil product: WTI grade crude oil, heating oil, gasoline and natural gas. Oil derivatives, on the other hand, come in all sorts of variations and combinations, yet can be generally classified into two broad groups in their exchange-based form: futures and options. At the same time, when traded over the counter, these instruments very frequently take the form of swaps (CFTC 2008).

Derivatives instrumentalise risk in such a way as to promote financialised accumulation, which abstracts from any linear relationship to underlying processes of real wealth creation. Under the guise of risk management, financial innovation has generated a plethora of derivative instruments which seem to simply mirror extant volatility, but in reality render volatility or variance a distinct traded asset. In turn, while justified by their ameliorative impact on uncertainty and role in optimising the capital allocation process, derivatives have profoundly altered a host of financial practices so that the financial sphere sits on top of the world economy attempting to profit despite, and indeed on the basis of, the vagaries of competition within it (Wigan 2008).

'Futures' are contracts whereby investors commit to buying or selling a specific volume of a commodity at a specific date in the future at a price that is determined at the initiation of the contract; the physical delivery of the product, which is required under the terms of the contract, can be avoided if the contract is sold before its expiration. 'Options' denote contracts whereby investors purchase the right, but not obligation, to buy a specific volume of a commodity for an agreed price level, within a specified period of time. Crucially, options do not bind investors to exercise this option; they just invest in the right to exercise their right at any point within their period of validity. This way, the exchange-based market of oil-based products provides the necessary tools for both long-term value investment, as well as risk hedging.

Utilising these products at their base, the unregistered, over-the-counter (OTC) market in which actors who trade bilaterally instead of through official exchanges are part of, is a very active platform for trading what is commonly known as 'Swap' contracts. These instruments are commonly seen to belong with the realms of the financial market, as they do not require the physical delivery of the product. Swaps can take different forms, either imposing a trade of the difference between the floating rate of oil and the average spot price over the period of the contract, or the exchange of an asset, based on the price of an index, for a similar asset with the intention of shifting risks or even adjusting maturity lengths. Thus, swaps offer a way to hedge risks while trading in futures and options as they are based on these products, yet avoiding the issue of product delivery, as well as going through the official exchanges where the trade will have to be registered.

From the historical perspective, commodity markets behaved, and performed, very differently from their financial counterpart before the early 2000s. This was because the factors that came into play in the commodity market were very different from the ones associated with the financial asset markets, which were predominantly affected by risk and other such factors. Although the 1990s witnessed increasing links between many financial institutions and underlying commodity markets, they were limited to strategies of portfolio diversification and risk hedging. Investment in the commodity markets was further limited by the requirement of physical storage and by the so-called 'prudent investor' rule, which, in defining investments in un-hedged futures contracts of commodities as speculative activity, discouraged the participation of many large financial institutions, such as pension funds. As a result, these markets remained long confined to their own circle of producers, consumers and corporations, as well as to the small class of financial actors and traditional speculators that specialised in their specific commodities.

REFERENCES

CFTC. (2008). *Staff report on commodity swap dealers & index traders with commission recommendations.* Washington, DC: Commodity Futures Trading Commission.

IEA. (2011). *Oil market report.* Paris: International Energy Agency.

Mabro, R. (2005). The international oil price regime: Origins, rationale and assessment. *Journal of Energy Literature*, 11(1), 3–20.

Wigan, D. (2008). A global political economy of derivatives – Risk, property and the artifice of indifference. Brighton: Brighton University.

CHAPTER 6

Oil Shocks as Barometers of the Financialisation Process

Abstract Having argued the multi-dimensional character of oil market financialisation as a process entailing the decoupling of asset markets and commodity markets, this chapter introduces the phenomenon of the oil shock as a barometer for clearly revealing and measuring the profound disconnect between market fundamentals and the actual market performance. Employing this approach marks the mechanisms of financialisation more visible than they would otherwise be under normal conditions, hence making it possible to lay bare the underlying dynamics of financialisation of the oil market.

Keywords Oil shocks · Asset markets · Commodity markets · Financialisation · Oil market

The previous chapter demonstrated the multi-dimensional character of oil market financialisation as a process entailing the decoupling of asset markets and commodity markets. The question that follows now is what would be the way of grasping the relevant stages of this process and creating the snapshots of those?

A market phenomenon that could clearly reveal the profound disconnect between market fundamentals and market's actual performance is that of the oil price shocks.

Although there are different approaches and definitions of oil shocks, the basic concept of this phenomenon refers to a sharp increase in the market price of oil. More specifically, according to Blanchard and Gali, in order for such an increase to qualify as a 'large oil shock' it has to be an 'episode involving a cumulative change in the (log) price of oil above 50 per cent, sustained for more than four quarters' (2008). These criteria match the shocks of 1973, 1979, 1999, 2002 and 2008.

Such price shocks can indeed serve as barometers of the financialisation process, demonstrating how – and the extent to which – the paper market performance seizes to reflect the fundamental performance of the underlying market commodity (owing to the rise of speculative and manipulative interests caused by financialisation). In this sense, if a shock acts as a window into the effects of financialisation, tracing a number of such consecutive crises would allow to benchmark the evolution of financialisation as such. At the same time, the effects of the oil shocks cannot be narrowed down to the analysis of transformations within the oil market itself. Rather, one has to look into the effects of such shocks and processes on the macroeconomic performance of an economy as a whole.

This is why the analysis of the relevant oil price shocks in this study will be placed in the context of US economic performance at that time.

Employing this approach will make the mechanisms of financialisation more visible than they would otherwise be under normal circumstances, hence making it possible to lay bare the underlying dynamics of financialisation of the oil market. In what follows, this study will proceed with looking into how the relationship between the oil shocks and the macroeconomic performance of an economy can be approached and analysed; and what does this analysis tell us about the process of oil market financialisation?

6.1 Oil Shocks and Macroeconomic Performance

The link between the oil price and the macroeconomic performance of an economy has been studied widely and confirmed by a number of studies such as Bruno and Sachs (1985), Hamilton (1996) and Blanchard and Galí (2008). On this note, Mork (1989) and Hooker (2002) contend that even though the relationship between the oil market and macroeconomic performance has changed and evolved through time, this is in part because

– and in spite – of the fact that both have evolved individually too; while Baumeister and Peersma (2009) argue that the oil market has changed along with global capacity utilisation rates, which have been operating in overcapacity from the mid to late 1980s.

Bernanke et al. (1997), in their *Systematic Monetary Policy and the Effect of Oil Price Shocks*, contend that the efficiency of national monetary policies has developed over time to make the macroeconomic structures more immune to changes in oil market volatility. This resulted from the gradual liberalisation of monetary policy institutions, especially central banks, as well as from a fuller understanding of the workings of the economy. This increased efficiency has also been attributed to an increase in the flexibility of labour and capital as well as to a change in the composition and role of the heavy industries.

In light of the above, it is important to stress that a linear analytical approach cannot fully capture the link between the oil market and macroeconomic performance. Rather, this link is better analysed as the evolution of relationship between two independently evolving, rather than constant, factors. The concepts of evolution, globalisation, financialisation, modernisation and even marketisation are a reality as much for the macroeconomy as for the oil and financial markets. The centrality of oil in debates about energy requirements, particularly in the West, has changed substantially since the 1970s. Therefore, this book approaches this study as a relationship of two evolving entities.

Turning to the effects of oil shocks, it is clear that they go far beyond the dimension of the oil market itself, having direct impact on the macroeconomy of both oil-exporting and oil-importing countries. When it comes to the latter, according to Roubini and Setser (2004), increased oil prices cause a subsequent increase in inflation rates and input costs and a reduction of real wages, of demand for non-oil products – for which less real income is now available – and of aggregate net investment levels. In addition to this, tax revenues decrease while the budget deficit likely increases accordingly, along with interest rates. Lower real wages and higher inflation lead to pressures for an increase in the level of nominal wages, which, combined with lower demand for non-oil products, create or exacerbate unemployment.

To that end, Lilien (1982) has formulated a 'dispersion hypothesis', based on the idea that oil price changes affect different economic sectors to different extents. More specifically, he explains that different sectors are dependent on different resources, so price changes in one or some of these resources will not affect all sectors equally. He also adds that, in the short run, the costs of readjusting essential resources have increased, which can result in a total loss of output. This, however, is rarely proportional to the contractions and expansions caused by the rise and fall of resource prices. This pattern becomes evident when observing the global levels of oil consumption and GDP against oil prices, and especially against the effects of the 2001 and 2008 oil shocks. The extent of these effects is determined by the expectations of an oil price increase; the more large and unexpected, the more it will take the markets by surprise to even more devastating effects. All these effects are all the more pronounced as financial investment, and speculation percolates through the oil market.

The negative effects of an increase in oil prices do not end there. On the external side of oil-importing economies, trade balance, in particular, will be distorted; this is due not only to the rising costs of oil imports, but also to the deteriorating health of the internal macroeconomic performance, which reduces productivity, competitiveness and, ultimately, exports. This, in turn, will negatively affect the balance of payments and exerts depreciation pressures on the exchange rate. As a result, oil imports will become even more expensive and this would lead to an international redistribution of wealth as increased transfers of value are made from importing countries to producing ones. What follows is increased budget deficits in the importing countries and increased budget surpluses in the producing ones, which intensifies all the negative effects described above: wider deficits and balance of payments, unless redressed by severe corrective policies, will exacerbate the levels of investment, productivity, prices and employment. The only way to prevent this economic race to the bottom is the introduction of heavy-handed, but prudent, monetary and fiscal policies.

As a result, a rise in the price level of oil will, at least in the short term, influence negatively most of the fundamental macroeconomic performance indicators of oil-importing economies. Simultaneously, the oil market has also the ability to influence the external aspect of importing countries' macroeconomies to a great extent, due to its influence on international exchange rates and trade deficits, inflated by the rising prices of an inelastic primary commodity (International Energy Agency 2004).

Clearly, the macroeconomic effects of oil shocks can provide valuable insights into the effects of the process of financialisation on the oil market. As the oil market went through the financialisation process, the behaviour and nature of the relevant market players operating within the financial segment of the market became no different to any other financial market as such. In other words, such actors constantly try to outguess the market by relying on past experiences and heuristics on the one hand, and equations and expectations based on economic fundamentals on the other. Therefore, as long as investors seek to increase their profits by the expected change in interest rates or by the returns on their investments, any change in the oil price level will cause financial adjustments as the values of the fundamentals change.

This has an impact on consumer confidence and expectations, which in turn influence financial markets through both the stock market and demand for credit and loans. The link between the oil market and the performance of the fundamental economic indicators of oil-importing countries spreads unevenly across the markets, while directly affecting any financial or economic decisions based on them. What is more, since many Western countries, such as the USA and the UK, have highly leveraged households with increased debt burdens which only rise in tandem with interest rates, savings and investment rates drop (Roubini and Setser 2004). Additionally, as the price of oil affects real household income, oil shocks will further reduce the percentage of it available for investments.

Recognising that oil shocks create ripple effects across the entire macro-economy of a country, the next part of the study offers an empirical analysis of the three periods of oil market's financialisation and the relevant macroeconomic context in the USA.

The transformation that the oil market underwent through these three phases of financialisation is demonstrated in Fig. 6.1. This figure reflects both, the respective oil price dynamics (WTI) and the activity of traders on New York Mercantile Exchange (NYMEX). One may immediately observe a dramatic increase in oil futures contracts and oil price volatility in line with the evolvement of the financialisation process. In what follows, these phenomena with be explained and mapped against both the structural changes within the financial segment of the oil market, as well as the effects of the oil price shocks on the macroeconomy of the country.

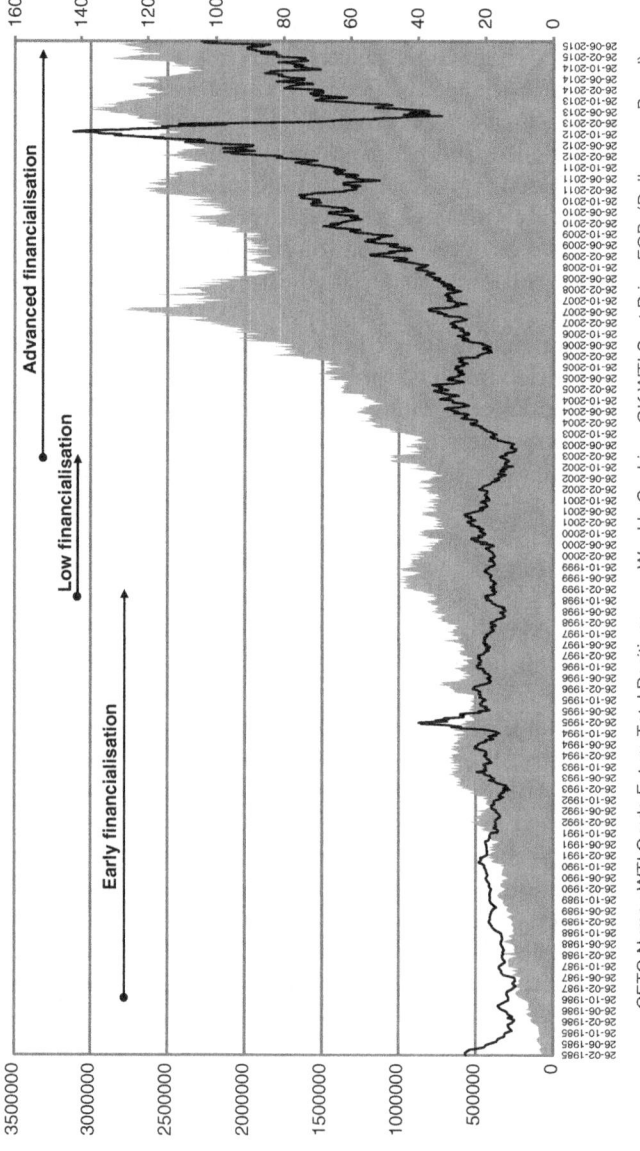

Fig. 6.1 Recorded oil futures NYMEX (1 contract = 1000 barrels) against weekly cushing, WTI Spot Price FOB (Dollars per Barrel) (Bloomberg Data 2015)

REFERENCES

Baumeister, C., & Peersman, G. (2009). Sources of the crude oil market volatility. *Bank of Canada and Ghent University*.

Bernanke, B., Mark, G., & Mark, W. (1997). Systematic monetary policy and the effects of oil shocks. *Brookings Papers on Economic Activity*, 91–157.

Blanchard, O., & Galí, J. (2008). The macroeconomic effects of oil price shocks: Why are the 2000s so different from the 1970s? In J. Galí & M. Gertler, *International dimensions of monetary policy*. Chicago: University of Chicago Press.

Bloomberg Data. (2015). *Data terminal*. London: Bloomberg. Accessed 14 October 2015.

Bruno, M., & Sachs, J. (1985). *Economics of worldwide stagflation*. Cambridge: Harvard University Press.

Hamilton, J. (1996). This is what happened to the oil price macroeconomic relationship. *Journal of Monetary Economics*, 38(2), 215–220.

Hooker, M. (2002). Are oil shocks inflationary? Asymmetric and nonlinear specifications versus changes in regime. *Journal of Money, Credit, and Banking*, 34(2), 540–561.

International Energy Agency. (2004). *Analysis of the impact of high oil prices on the global economy*. Paris: International Energy Agency.

Lilien, D. (1982). Sectoral shifts and cyclical unemployment. *Journal of Political Economy*, 90(4), 777–793.

Mork, K. (1989). Oil and the macroeconomy when prices go up and down: An extension of Hamilton's results. *Journal of Political Economy*, 97(3), 740–744.

Roubini, N., & Setser, B. (2004). The effects of the recent oil price shock on the U.S. and global economy, Stern School of Business NYU and global economic governance programme. University College Oxford.

CHAPTER 7

The Three Phases of Oil Financialisation: Early Financialisation (1980–1999)

Abstract This chapter utilises the combination of a historical review with a macroeconomic and financial analysis, in order to create a snapshot of the status of the financialisation of the oil market during the period of early financialisation. Developments such as those of the turbulence in the Middle East, oil embargoes, the US macroeconomic performance and the introduction of the first commodity indexes are studied closely before reaching the conclusion that the emergence of the oil-based financial tools created the necessary prerequisites for the participation of a new type of investor in the oil market, one with non-traditional financial interests, who however was still at an infant stage during the period of focus of this chapter.

Keywords Phases of financialisation · Oil market · Investment · Financial interests

7.1 Macroeconomic Background

The US political–economic situation after the 1979 oil shock was recovering fast: the USA was in a very strong international position under President George Bush; the US economy was booming, with growth rates reaching 4.5 per cent in 1988; the unemployment rate was stabilised between 5 and 5.5 per cent; and the inflation rate rarely exceeded the barrier of 4 per cent. However, the Federal Budget was in a very difficult position as this boom had been underwritten to a large extent by government expenditure.

For this reason, during the late 1980s the Federal Reserve was attempting to guide the US economy to a 'soft landing' by gradually reducing its budget deficit (Cooper 1992, p. 152). Their aim was to reduce growth levels to a sustainable 2/2.5 per cent, while avoiding a further slowdown that at the time would have likely caused a recession. Up until the beginning of the 1990s, the contractionary monetary policy adopted by the Federal Reserve seemed largely successful. At the same time, the US President had agreed to hold talks to reduce the levels of government expenditure.

Nevertheless, during these discussions, events from the Middle East got in the way of this path towards economic readjustment and stabilisation. Iraq invaded Kuwait on August 1990 as they announced that Kuwait and Saudi Arabia were committing an act of war by increasing their levels of oil production, which had led to a reduction in the agreed price of oil from $18 a barrel to $7 per barrel (Hybel 1993).

The USA made it clear that Iraq's actions would not be tolerated (Cooper 1992) and deployed forces on Saudi Arabia in order to expel the Iraqi forces out of Kuwait. It has been calculated that the military 'bonus' expenses of this war surpassed the $100 billion mark over the period from 1991 to 1995 (Korb 1992). What is more, the USA not only isolated Iraq's and Kuwait's accounts on foreign soil, but also introduced a full embargo effective to all their products, including oil. Considering that the states of Iraq and Kuwait held 20 per cent of the total world oil production at the time, this embargo clearly caused great disruption to the oil market equilibrium of supply and demand. Any supply restriction of this magnitude leads to an increase in the price level, especially since the levels of demand are inelastic given the nature of the good and the speed of adjustment of the market. This figure becomes even more significant when taking into account that the US economy imported a higher amount of its oil from the Middle East. This restriction of supply levels was therefore reflected on the price level, which more than doubled within months of the beginning of the conflict.

Even though strategic oil reserves had already been built, especially in the USA, after the events of the 1970s, they were not released due to the political implications of such action. As expected, the spare-capacity levels of the OPEC countries' oil production dropped significantly during this period, reflecting the turbulence that these conflicts had created within the Middle East.

As a result, the embargo of Iraqi and Kuwaiti oil, along with the disruption of the market equilibrium caused by the military operations, led to what is known as the third oil shock. The price of oil more than doubled, from $15 to $33 per barrel. What sets this particular oil shock apart from the other ones studied in this book is mainly its duration; this shock was short-lived, as can be characterised more accurately as a spike than as a large oil shock. However, even though the actual hike in the price level was short-lived, its effects in the US economy were extensive. The study of these effects may shed some light on the functions of the oil market and on its links to the financial and macroeconomic performance.

The historical context behind this oil spike is significant. At the time, in conjuncture with the government budget reviews, the Federal Reserve was attempting to slowly reduce growth and inflation to more sustainable levels without running the risk of sliding into a recession. This process, already in place before the spike hit, had been largely successful until the Gulf War and the oil price spike. Prior to mid-1990, US interest rates had increased as a result of the Federal Reserve attempt to reduce growth and inflation rates, which stood now at around 2.5 and 3 per cent, respectively. With unemployment stably below 5 per cent, the US economy, and its contractionary monetary policy, was indeed faring well.

7.2 The Oil Price Spike Effects

The situation described above changed dramatically after the first half of the year 1990, for the oil price spike and the possibility of a war in the Gulf had direct and extensive effects on the markets. The increased oil prices, along with the fear that the military activity would affect oil prices further, gave rise to an increasing wave of negative expectations across the US markets. In addition, the refusal by the US government to release its oil reserves hardly helped to reduce the uncertainty and negative expectations in the markets, which were desperate for some measure of reassurance.

By the middle of 1990, and precisely on 2 August 1990, all the basic fundamental economic indicators had been destabilised. First, there was both a noticeable acceleration of the inflation rate, which was met by a decrease in interest rates, and a reduction of the real growth level, also reflected in the unemployment rates. While the inflation and growth levels decreased after the end of the Gulf War, unemployment and interest rates did not recover until the second half of 1992. At the same time, while a revised budget was being agreed, the deficit continued rising, reaching $279 billion in 1991, and was not helped by a devaluated exchange rate.

The USA was faced with increased expenditure levels, and at the same time was being locked in the position where of having to purchase more expensive oil in a devalued currency, which raised its deficit level even further. The loss of value of the US dollar was evident throughout the 1990s. The actual loss of value was more than 10 per cent. It is also worth noting that, though oil prices normalised in 1991, the unemployment level and the budget deficit did not start recovering before the late 1992.

At the consumer level, US personal income and consumption only decelerated in 1991 and was virtually unaffected in 1993. This is because the 1991 crisis affected both personal income and public expectations, which stabilised savings and reduced spending. In contrast, the 1993 reduction of personal income was short-lived and was not accompanied by the expectation of a crisis, thus the level of consumption was spared.

The financial markets were affected by these events as much as the macro-economy. The volume of layoffs in Wall Street increased during this period as a result of the destabilisation of the banking system (Stewart 2008). People and firms were defaulting on loan payments, while the US real estate sector was failing as real estate prices went down. However, the stock market performance during this period suffered from a substantial deceleration. Even though this financial crisis is often characterised as a 'mild' one, the drop of the stock SPX index from above US$360 to below US$300 is symptomatic of a crash that, albeit less than a year long, extensively affected its financial performance while not necessarily being a product of its doing. Therefore, even though the financial sector did not directly affect the events of 1991, it can be argued that the events of 1991 did directly affect the financial sector.

7.3 Financial Dimension of Oil Markets: The Emergence of Oil Futures and Swap Investments

As the oil market was not yet widely established as a product tradable in the financial markets, this deceleration can only be attributed to its links with the macroeconomic performance as well as to the increased drop in the consumer confidence levels of this period. Although the IPE released the first oil futures contract in 1981, the first Brent Crude future was actually launched in 1988, and the market has experienced continuous growth virtually every year ever since.

The introduction of those contracts allowed market participants to purchase a risk-hedging product that could provide insurance on the investments they were involved with in the physical oil production market. This is because these futures contracts had the ability to provide an agreed and guaranteed forward price valuation for oil, which could be used as a guide for their investments, as well as for calculating future costs and returns. This future price valuation was not necessarily the optimal one as factors of risk, expectations and uncertainty were always involved; but they did provide with a measurable and guaranteed value. As a result, many market participants, such as producers, heavy consumers, refiners or marketers of oil, entered the market to insure their investments.

In response to the need for a simplified investment option for oil markets, with lower risks and no obligation of physical delivery, Goldman Sachs seized the opportunity to build an asset class based on commodities. This asset class, the S&P Goldman Sachs Commodity Index (GSCI), employed a combination of various commodity futures and options, with a significant large proportion of it reserved to oil. This meant that investors were now able to invest in oil without the problem of commodity storage, thanks to the fact that these indexes operated on the value of the futures contracts with a minimum expiration period of one month, and more specifically, on the principle of 'rolling' from one contract to the next. This way, investors were exempted from taking physical possession of the commodity and their investments operated as if holding a long position on a futures contract (Silvennoinen and Terosvirta 2009). An additional advantage was that the index was based on a number of different commodities, so investments also enjoyed risk diversification across all these commodity markets.

The first trades on the S&P GSCI took place in 1992, but it was not long before other indexers followed suit (Wray 2009). The two most popular indexes, the S&P GSCI and the Dow Jones–AIG Commodity Index (DJ-AIG), have recently been joined by a number of smaller indexes, such as the Rogers International and the Deutsche Bank Liquid Commodity Index, which vary considerably in their specific characteristics (Tang and Xiong 2009). All these commodity indexes function the same way as equity indexes (e.g. S&P 500) as their value is calculated from a specific basket of commodities, where each commodity differs in weight.

The commodity markets, being positively correlated to the rate of inflation, provide a very useful portfolio diversification tool compared to other financial assets. While being marketed as future contracts, commodities are part of the basket of goods that is used to calculate inflation rate levels; their value is, therefore, determined by the same expectations that regulate expected inflation levels (UNCTAD 2009). Furthermore, as most commodities are traded only in US dollar values, investments in the commodities market provide a very effective hedging strategy against the exchange rate of the US dollar, while, at the same time, the value of the US dollar affects the international demand for said commodities (Clapp 2009). The fundamental characteristics of all those oil-based financial products made accessible by the indexes marked them as effective hedging investment options. Nevertheless, the regulatory and logistical barriers of the market still made such investments either unappealing or inaccessible to many large financial institutions.

After 1992, another type of investment became available in the oil market in the form of the swap investments, which are considered as being highly speculative investment products (CFTC). The volume of non-commercial positions started increasing gradually and, at the same time, so did its levels of volatility. Commercial positions, too, experienced an unprecedented increase in their volumes over this period. While up until 1990 they were struggling to reach 250,000, in 1999 they exceeded 800,000. It is also interesting to note that the volumes of long and short positions are similar in both commercial and non-commercial investments after the development of the indexes.

The findings above suggest that the emergence of oil-based indexes created the necessary prerequisites for the participation of a new type of investor in the oil market with non-traditional financial interests. They also suggest that, while the purely financial investments quickly caught up with the rest of the market in terms of overall volume, the levels of the traditional long and commercial investments increased as well. Therefore, this period witnessed a general increase in the appeal of the oil market as an investment opportunity, with commercial positions still holding a dominant share in the market.

The same argument can be made when it comes to the data on oil futures prices relative to their maturity – typically a breakdown of all the one-to-five-year futures contracts. The pricing of oil products with different maturity is defined by the current price of oil as well as by the expectations on its future price in order to calculate their potential returns, and thus pricing. When expectations for future returns are positive, the price difference among the one-to-five-year maturity contracts should increase, as higher expected returns translate to higher present prices. In this case, the factor of risk is also included in the pricing of these products, as shorter maturity products are less risky than long maturity ones.

To conclude, when the expectations for the future returns are negative or uncertain, the price difference between the different futures contracts should decrease, as longer maturities imply higher risks. In the post-1995 period, oil started to be considered an increasingly riskier of an investment opportunity, heavily dependent on the expectations of market players. Both of these features of the market are rooted in the emergence of a new type of a market segment and market actors with purely financial motives. This way, short, non-commercial and risky investments started becoming more popular among purely financially driven investors who were mostly positioned outside the oil market structure. For this reason, this period is viewed as the period of 'early financialisation': throughout its duration, it laid the institutional foundations that allowed the process of financialisation to penetrate deeper into the structure of the oil market.

REFERENCES

Clapp, J. (2009). Food price volatility and the vulnerability in the global south: Considering the global economic context. *Third World Quarterly*, Routledge.

Cooper, R. (1992). The Middle East and the world economy. In J. Nye & R. Smith *After the storm, lessons from the Gulf War*. London: Madison Books.

Hybel, R. (1993). *Power over rationality, the Bush administration and the Gulf crisis*. New York: State University Press.

Korb, L. (1992). The impact of the Persian Gulf War on military budgets and force structure. In J. Nye & R. Smith *After the storm, lessons from the Gulf War*. London: Madison Books.

Silvennoinen, A., & Terosvirta, T. (2009). Modelling multivariate autoregressive conditional heteroskedasticity with the double smooth transition conditional correlation GARCH model. *Journal of Financial Econometrics*, 7(4), 373–411.

Stewart, J. (2008), April 1. Lessons From the 1990–1991 Recession. *smartmoney.com*.

Tang, K., & Xiong, W. (2009). Index investing and the financialization of commodities. *Princeton University*.

UNCTAD. (2009). *Trade and development report 2009*. New York: United Nations Publications.

Wray, L. (2009). The rise and fall of money manager capitalism: A Minskian approach. *Cambridge Journal of Economics*, 33(4), 807–828.

CHAPTER 8

The Three Phases of Oil Financialisation: Low Financialisation (1999–2002)

Abstract This chapter utilises the combination of a historical review with a macroeconomic and financial analysis in order to create a snapshot of the status of the financialisation of the oil market during the period of low financialisation. Developments such as those of the dot-com bubble, the stock market crash, the election of Hugo Chávez as the president of Venezuela and his bid to take control of the oil market, the introduction of Internet-based trading of oil futures and options, as well as the Commodity Futures Modernization Act regulation and the abolition of the 'prudent investor' rule opened the doors to new investors entering the oil-based financial market. These new investors were now substantial enough to be able to influence its performance via an increased amount of capital in the oil market.

Keywords Historic finance · Financialisation · Oil market · Stock market crash · Investment

8.1 Macroeconomic Background

The final years of the previous millennium were characterised by unprecedented growth in most Western economies, and especially in the USA, up to the year 2000. The economic expansion of USA had lasted for almost a decade since the end of the 1991 crisis and rates; in fact, it is considered as the longest registered expansionary period experienced by any industrialised country (Arestis and Karakitsos 2004). The extent and strength of this expansion can be demonstrated by reference to the US GDP growth levels, which by the end of 1999 had reached 7 per cent. The inflation rate, kept at low, stable levels, averaged 2.5 per cent during the late 1990s, while the unemployment level reached 3.9 per cent in 2000 – its lowest level since the 1970s. According to Temple (2002) even poverty and wage inequality levels had decreased substantially.

It did not take long for this extraordinary economic expansion to spill over into the stock markets. From 1995 to 1999, the US stock exchange Standards and Poor's Composite Index (S&P 500) produced a total return of over 20 per cent per annum. The best performers in the stock markets during this period were the stocks of companies in the sector of media, telecommunications and especially technology. According to Rima (2002), the stock prices of some of these companies rose by more than 1000 per cent from their initial public offering (IPO) price. This development created the feeling in the markets that a 'new economy' had emerged after the previous crisis, which was more resilient to shocks, thanks to new rules of operation (Stiglitz 2003).

This was the result of the transition from a mainly manufacturing-based Western capitalist system into a service- and finance-based economy that was proving some past economic truths, such as the Phillips Curve theory, obsolete. The development of novel and advanced technologies throughout this decade had given firms from across most sectors the opportunity to claim that the new technologies had been so transformational that the limitations of the past were quite simply no longer applicable. This led to a sustained increase in stock values across all of these sectors. In many cases these stock gains were completely disconnected from the actual economic value or performance of the firms they represented.

Attempting to capitalise on these positive valuations, an increasing number of very successful IPOs were released during this period (Henwood 2005). This allowed many firms to raise significant amounts of capital from the markets, which they would have otherwise struggled to raise in the form of debt based on their fundamentals alone. Unemployment and inflation were very low and output levels remained steady and positive, thus creating an optimal economic environment. According to Nesvetailova (2007), however, the underlying economy gave signs of a marked discrepancy between output growth and the increase in the share of profitability. This argument is also shared by actors directly involved in these markets. In a statement published by the trading body of Total Oil, "several times a year, estimates of market prices on key energy indexes are out of line with the experiences of the day" (Makan et al. 2013).

As stock returns rose, unprecedented profits were being made in the stock markets, and especially in the technology sector, in such a loud trend that it quickly became common knowledge. Not surprisingly, massive inflows of capital entered the stock market in the direction of those shares, which generated what is now known as the dot-com bubble (Krugman 2009). Blind to the sustainability of this situation, everyone seemed to throw all hesitation to the wind for the sake of more profit. Both per capita income and consumption levels experienced an unprecedented increase in the period leading up to 2001. By earning and spending more, all the economic and financial participants in the US economy contributed to a positive reinforcement cycle – and to the inflation of the bubble.

According to Shiller (Shiller 2005), the dot-com bubble, and the euphoria that surrounded it, had created a mania. On top of this, financial innovation had started to peak during 2000 and 2001, when several US energy and electric utility companies, the most famous of them being Enron, started to engage in derivatives. Additionally, this positive environment was also sustained by the lowest price levels in the oil market since the 1970s, which reached just over $10 per barrel after the 1992 oil price spike. At the same time, George W Bush Jr. won the 2000 US presidential elections and, upon taking his post, set out to fulfil his electoral promises of lower taxation.

The Federal Reserve started to express concern in 1999 that the growth levels experienced by the US economy would be unsustainable in terms of future inflationary consequences, and that 'corrosive inflation' was soon likely to set in (Greenspan 2008). High growth levels indicated that the expansionary leg of the cycle was peaking out and that inflationary pressure would soon take over. The Federal Reserve resolved to increase interest rates in order to slow down growth and steer clear of any inflationary pressures, thus prolonging the boom. Their initial increase was not met by the expected reaction of the markets, but they believed that this reaction was typical of economic actors who were adjusting to the new conditions. As a result, the Federal Reserve decided to raise the rate even higher, up to 6.5 per cent.

However, the reaction of the markets to this last interest rate increase was not the one desired by the Federal Reserve, as the economy started decelerating sharply and continuously. The US economy fell into a recession in the beginning of 2001 and the confidence that had been built up in relation to the 'new economy' evaporated as the dot-com bubble started to deflate, creating huge debts and defaults in its wake (Rima 2002). The stock markets crashed, leading to a loss of about 40 per cent of their average value. This turn of events alarmed the Federal Reserve to the point of veering towards expansionary policies of low interest rates in an attempt to put an end to the recession. Between 2001 and 2003, the Federal Reserve announced thirteen reductions in the interest rate, along with the introduction of additional expansionary fiscal measures on the part of the government. The budget surplus achieved towards the end of this decade of expansion was now very quickly turning into a deficit.

At this point, production slowdown took place across the US economy; most firms announced lower corporate profit estimates; unemployment pressures started to creep in and inventories started to pile up due to an increasing deficit in demand (Greenspan 2008). Consumer confidence was now severely hit; the levels of industrial production suffered as a result and fell by 5 per cent. The manufacturing sector attempted to respond to these effects by cutting down production levels, shedding labour and minimising investment levels. Even with these measures in place, the corporate sector was unable to witness any reduction in the levels of accumulated stocks; with the inventory-to-sale ratio increasing and sales continuing to fall, production was being cut across most sectors of the economy (Arestis and Karakitsos 2004). For this reason, this recession is also referred to as 'the inventory crisis'. Defaults became a worldwide phenomenon: only in 2001, 216 companies defaulted on US$116 billion of debt, a trend that continued in 2002. What is more, even 25 large US companies filed for insolvency (Nesvetailova 2007), which proved that this crisis did not only affect dot-com firms but also small-to-medium enterprises.

At the same time as these events were taking place in the US economy, developments in the oil market were affecting the international economies. Hugo Chávez was elected in 1998 as the president of Venezuela. In the years preceding Chávez election, OPEC had lost most of its former glory, as members frequently ignored the agreed quotas with the result of a dive in the price of oil and the economic boom of the 1990s. In 1999, within the first one hundred days in power, Chávez succeeded in taking control of the oil market and increasing the price level by $5.19 per barrel. Later, in the same year, he achieved further increases through supply restrictions (Rodríguez 2002).

In 2000, the price of oil reached $30 dollars per barrel, forcing the release of a major portion of the US Strategic Petroleum Reserves in an effort to fend off the negative effects of this increase. This decision helped to temporarily lower the price level of oil in the US economy; however, the events that followed cancelled out the positive effects of this release. The combination of high oil prices, increased interest-rate levels and the gradual descent of the budget surplus into a deficit put an end to the euphoria of the previous decade as the economy fell into a crisis.

The US economy, being the biggest importer of oil internationally, is very vulnerable in its macroeconomic performance to changes in oil price. In the year 2000, according to the EIA database, the USA was importing 11,459 thousand barrels of oil per day, which, considering the $20 increase in the price of crude oil from 1999 to 2000, accounted for approximately $230 million more paid every day by the USA for their crude oil import needs. Higher oil prices lead to higher budget deficits.

8.2 Technology and Regulation: The Three Factors that Changed the Oil Market

One of the turning points for the development and financialisation of the oil market in that period was certainly the penetration of the Internet in its structure. In 1999, oil futures and options became available for trade to any individual. Entry to the market had never been easier: trading oil-based products now only required a subscription to a trading website. This was a very important development for the financialisation of the oil market as, up to that point, it was very hard and almost impossible for an individual to trade directly in oil-based financial products. Once the latter became accessible to individuals across the globe, the market of these products has widened on both a geographical and demographic level, thus further intensifying the financial segment of the oil market.

A second development that occurred during this period and allowed for the intensification of the financialisation effects in the oil market was the Commodity Futures Modernization Act of 2000 (CFMA). The CFMA of 2000 was approved by the US Congress on 15 December 2000, and signed into law six days later by the former US President Clinton. This act effectively deregulated all the new oil-based financial products, such as the oil futures, options and swaps. It was achieved by marking a number of OTC derivatives off the Commodities Futures Trade Commission (CFTC) jurisdiction and therefore allowing financial actors to take advantage of this loophole and trade these products through the OTC markets, while being out of the reach of CFMA regulations.

This development affected the financial aspect of the oil market in two distinct ways. First, it provided investors with room for financial innovation as well as a loophole to trade in this market free from regulations. In other words, in an attempt to provide more stability in the market, this act allowed for a larger volume of capital to enter the market and to be traded more actively. The second effect on the market is a consequence of this development. As OTC trades are not recorded the same way as traditional trades, and therefore cannot be accurately quantified, the increased activity in the OTC markets after the introduction of this act cannot be assessed precisely. Hence, any calculation of the oil futures market after 2000 overlooks the parallel OTC market, which in practice is entirely discounted in their values.

The third factor was the abolition of the 'prudent investor' rule, which was replaced by the Uniform Prudent Investor Act (UPIA) in 1994. This was a legal doctrine that enjoined financial actors who invested with third-party funds to consider the needs of their beneficiaries and to act in a prudent and diligent way. In other words, it required them to avoid excessive risk in their investments by adopting very conservative investment strategies. What is more, it stated that the final result did not justify in any case the original intent of the investment (Einloth 2009). This meant that financial institutions, such as pension funds, hedge funds and banks, could not engage in the oil futures market as this would amount to investing funds not belonging to them in a market, known for its high-risk character.

Under the UPIA approach, investors could focus on the portfolio as a whole, and the interplay between risk and reward, without any restrictions or distinct rules when it comes to safe or risky investments. Investing institutions could now establish the level of risk that they are willing to take and then map out an investment strategy for their portfolio, with the only provision that they should be able to explain the rationality of the strategy and the prudence of each investment as it relates to the whole portfolio. The abolition of the 'prudent investor' rule also opened the doors to a wide variety of investment products that had been introduced, or come into the mainstream, since 1959. If in 1959 there were 155 mutual funds with nearly $16 billion in assets, these figures had already grown to 10,725 and $6.9 trillion, respectively, by the year 2000 (CDA/ Wiesenberger). As a result of this development, financial institutions such as pension funds and hedge funds, now controlling an increased amount of capital in the markets, were able to add oil-based financial products to their portfolios and thereby increase the inflow of capital in the market as well as the number and influence of its participants (Tang and Xiong 2009).

REFERENCES

Arestis, P., & Karakitsos, E. (2004). *The post bubble US economy: Implications for financial markets and the economy*. London: Palgrave Macmillan.

Einloth, J. (2009). Speculation and recent volatility in the price of oil, division of insurance and research. *Federal Deposit Insurance Corporation*.

Greenspan, A. (2008). *The age of turbulence*. New York: Penguin Books.

Henwood, D. (2005). *After the new economy*. London: The New Press.

Krugman, P. (2009, July 8). Oil Speculation. *Nytimes.Com*.

Makan, A., Blas, J., & Spiegel, P. (2013), May 14. European commission raids oil groups over price benchmarks. *Financial Times Online*.

Nesvetailova, A. (2007). *Fragile finance, debt, speculation and crisis in the age of global credit*. Hampshire: Palgrave Macmillan.

Rima, I. (2002). Venture capitalist financing: Contemporary foundations for Minsky's Wall Street perspective. *Journal of Economic Issues*, 36(2), 407–414.

Rodríguez, A. (2002). *Saramago: Soy Un Comunista hormonal*. Venezuela: Capital Intelectual.

Shiller, R. (2005). *Irrational exuberance*. New York: Princeton University Press.

Stiglitz, J. (2003). *The roaring nineties – A new history of the world's most prosperous decade*. London: W. W. Norton & Company.

Tang, K., & Xiong, W. (2009). Index investing and the financialization of commodities. *Princeton University*.

Temple, J. (2002). An assessment of the new economy. *CEPR Discussion Papers*.

CHAPTER 9

The Three Phases of Oil Financialisation: Advanced Financialisation (2002–2015)

Abstract This chapter utilises the combination of a historical review with a macroeconomic and financial analysis, in order to create a snapshot of the status of the financialisation of the oil market during the period of advanced financialisation. Developments such as those of the US housing bubble, the subsequent financial and economic crisis, the September 11 attack and the war in Iraq which led to the oil price spike of 2008 are argued in this chapter to have driven a rally in speculative activity deriving from the financial actors who were now able to enter the market freely, and in large numbers, after the regulatory developments of the previous period. This thus demonstrates clearly the advancement of the financialisation process in the oil market, which had now reached the level whereby the capital from other segments of the market could flow freely into that of oil. By now, both the actors and the modus operandi of the financial element of the oil market had reached its maturity and had become a rightful part of the international financial markets system.

Keywords Historic finance · Economic crisis · Speculation · Oil market · Financialisation

9.1 Macroeconomic Background

The common understanding of the 2008 crisis is that in 2007 a bubble in the US housing market broke as a result of the proliferation of subprime mortgages, which were being used in the formation of synthetic financial products, and caused the collapse of the financial market. The reduced capacity of banks to absorb this burst was mainly due to the kind of insufficient funds, risky activities and enormous short-term lending that they had hitherto engaged with. This unexpected shock in the international financial system led to a crisis of confidence, mainly towards the credit organisations that used short-term lending as their main source of financing and, in doing so, left themselves open to a liquidity crisis (Akerlof and Shiller 2009).

It did not take long for the effects of the financial crisis to seep into the macroeconomy. Initially, the unwillingness of credit institutions to grant loans, on account of their degree of underperformance due to increased mortgage default rates, created liquidity problems across the markets. Second, the crisis caused the evaporation of a great share of the international household and corporate wealth, as many people lost large parts of their investment capital either through institutional defaults, the crash in the financial markets or the loss of real-estate investment value. Following from this, the levels of international trade collapsed, as the lower levels of consumption in the most widely affected economies – also the world's biggest import economies, that is the EU and the USA – led to increasing unemployment pressures (Akerlof and Shiller 2009).

No reaction from the international stock markets was forthcoming until the early 2008, which seemed to vindicate the argument that a decoupling of the international financial markets from the US economy had indeed taken place. Nevertheless, the downward turn of the international stock markets from January 2008, which lasted for about fifteen months, proved otherwise. Notably, the fall of financial corporations, which was the central market of this crisis, was much wider than that of all the other sectors. In 2009, the international economies were faced with the biggest recession of the post-war period. It is remarkable that up to 2007 economists and analysts were proud for achieving sustainable high rhythms of world growth, combined with low inflation.

Demand for employees in the financial sector during the run up to 2007/2008 increased by more than 15 per cent in order to cover the growth of its activities. This rise, not surprisingly, ends at the beginning of 2008 when the levels of employment in this sector plummet to below their pre-bubble-emergence levels. At the same time, even though the crisis hit the financial sector and the employment levels of this sector dropped significantly, the percentage of the financial sector in the total annual GNP composition remained stable and even increased during this crisis because the sector was flexible enough to find ways to remain profitable.

Cheap money, excessive liquidity, and very low real interest rates shaped international imbalances through saving and investment activities and acted as the drivers of this crisis (Portes et al. 2009). These imbalances were fuelled by the high deficit of the USA and low real interest rates, which inflated the bubbles that existed in the real estate market and that of the complex financial products. The housing market bubble in the USA emerged in 2001, when the real estate prices began to rise, yet the market was left free to expand uncontrollably – a phenomenon puzzling a large number of analysts (Hardouvelis and Stamatiou 2009).

In addition to the expansionary monetary policy, the USA was engaged in a very costly war in the Middle East at the time and maintained a large number of troops in the region. The increased budget deficit observed above in relation to the previous decade was maintained through the high costs associated with this war, which however boosted the productivity of the US automotive and military industries and reduced the unemployment levels.

Against this background, in 2007 things started to change. A deceleration in the growth of the housing-market prices started in mid-2006, with prices falling in 2007. This drop led to delays in mortgage payments and many households leaving properties that were suddenly worth less than the mortgages they commanded. This intensified the effects of the crash as properties began flooding the market, pushing their prices even lower. The reduction of household wealth, along with the unfavourable financial environment, led to even more defaults in the remaining mortgages, thus further decreasing the available funds on which many advanced financial products were based.

In February 2007, the then new Federal Reserve Chairman Ben Bernanke stated that the growing number of defaults in the payments of mortgages would not affect the US economy seriously. However, it was not long before trouble reached the major Wall Street firms. To make things worse, in September Northern Rock, a UK bank, was irreparably hit by the crisis and asked for assistance from the Bank of England. Depositors, fearing the loss of their funds, withdrew one billion pounds in what was soon turning into a bank run before the UK government intervened with a guarantee of depositor savings.

In March, Bear Stearns, the fifth largest investment bank in the USA, collapsed. A few weeks later, Fannie Mae and Freddie Mac were effectively nationalised by the US Treasury. At this point, the International Monetary Fund (IMF) announced that the potential cost of this crisis could surpass one trillion dollars, claiming that a spillover effect was taking place from the subprime mortgage markets to the other sectors of the economy and the financial markets. In September, Lehman Brothers, one of the largest and most successful financial services firms, filed for bankruptcy after the US government denied it a bail-out to avoid, giving the wrong signals to the markets. The stock markets plummeted, as confidence was lost, while investors and partners of Lehman Brothers lost their funds.

The collapse of Lehman Brothers, even though it was not the first firm to get hit by the crisis, was particularly significant because it demonstrated that even firms deemed 'too big to fail' were, in fact, failing. Large banking institutions were struggling to survive in such a negative financial environment, but were still in a better position than many unregulated institutions, such as hedge funds, which had engaged in high-risk investments without any form of restriction and were now suffering extensively as a result of it. Negative expectations reduced liquidity, faltering demand levels, crashing markets, and the strict application of the no-bailout policy by the US government meant that there was no alternative for many such institutions other than default.

The US consumer confidence index dropped to the record low level of 28 points. This was the lowest value that the consumer confidence index had reached since the Conference Board began tracking consumer sentiment in 1967. Due to the lack of confidence and fear of increased risks, investors retracted their capital from developing and high-risk countries, thus spreading the effects of the crisis to countries with undeveloped financial markets. Against this background, on Monday 1 December 2008, the US economy officially entered an economic recession, after the contractionary path started in December 2007.

The financial crisis and its effects on the macroeconomic performance also affected the levels of international trade. According to Freund (2009), during the four post-war recessions, international trade dropped by 4.8 times more that the world Gross National Product (GNP). The evidence available for the first half of 2009 illustrates a much bigger drop, one that exceeds 14 per cent of the world GNP. It was the composition of the international trade itself and the globalisation effect that arguably constituted the main causes for this drop (Baldwin 2009). The recession, along with the feelings of uncertainty and insecurity, stirred a reduction in consumption and, in turn, production.

Against this background, after the attacks on September 11, the USA and a few of its allies, initiated a military attack against Iraq. Even though this was expected to be a short and a cheap war (Stiglitz 2008), it turned out not to be the case. Although there is no official figure, there are many credible estimates of the total costs of this war. Joseph Stiglitz argued that by the end of the Bush administration, the total cost of the military activities in Iraq and Afghanistan, taking under consideration the cost of the cumulative interest on lending that was employed to ensure their funding, surpassed the US$3 trillion figure. Figures of this sort are particularly striking considering that the White House expected the war to cost between US$100 and US$200 billion while the Defence Secretary Donald H. Rumsfeld insisted on no more than US$50–60 billion budget. On the one hand, there were some benefits of this increased spending for the US economy, in particular increased output, profitability and demand for labour. On the other hand, the US budget deficit did not enjoy the same positive performance during this period. Increased military spending, combined with low taxation levels, led to a devastating increase of the budget deficit levels.

9.2 The 2008 Oil Price Shock and the Role of Speculative Financial Activity

Adding to the increased deficit level, the oil price spike at the time exerted pressure on the US dollar, while naturally leading to an increase in the US expenditure level. This oil price spike also resulted in the increase of the transportation and production costs and as a consequence, inflationary pressures. The oil market experienced an unprecedented oil shock, which saw more than a 400 per cent increase in price levels, which shot up from $30 per barrel in 2003 to more than $140 per barrel in 2008.

The direct effects of the conflict in the Middle East on the oil supply and on the oil shock that took place during this period have been a subject to an extensive academic debate. The war in Iraq did disturb the oil export levels, with Iraq being one of the biggest oil suppliers to the world. However, the extent of this oil shock was much larger in magnitude compared to the suggested restriction in supply levels. In 2002, the average trading volume of oil futures was four times greater than the volume of demand for the actual physical product; in 2008, and the beginning of 2009, it was fifteen times greater (Khan, 2009).

Michael Masters and White (2008) contended that the oil price spike of 2007/2008 was caused by the investment decisions of actors who had no interest in the physical product, but only in its value as a financial asset.

He goes on to claim that by March 2008, the oil futures index had risen by a quarter of a trillion dollars, and that the typical financial actor strategy that was employed was about taking long positions in short-term futures contracts and keep selling them days prior to their expiration, before moving onto the following one. Therefore, what Masters proposes is that the financialisation of the oil market allowed for the futures and spot market to be distorted by speculative financial activity.

A report published by the OECD (2010) on speculation in commodity futures market uses the volume of open interest in the futures markets to identify the increase in activity during the period between 2006 and 2009, which is argued to be a result of the commodity indexes. Tokic (2011) calls for a study that should include all the participants of the crude oil futures markets to determine the existence and significance of speculation during the 2008 oil bubble. In his later analysis, Tokic identifies a reduction in the net short positions held by commercial hedgers leading to the peak of the oil bubble in 2008, and therefore concludes that positive feedback trading through short covering might have been a significant contributor to the increase of oil prices and the inflation of the bubble.

The 2008 UK Cabinet Office Report argues that, despite the lack of concrete evidence for the existence of speculation and for its role in the 2008 spike, some indicators of such activity do exist. The report claims that it is possible to trace the difference in the investment behaviour between the financial and the traditional investors and, therefore, to explain the behaviour of the market. This is because financial investors used oil as a hedging mechanism in their investment portfolios, which means that the similarity of their motives and information made them act in a similar way. They mostly hold long positions, not being influenced by short-run changes and market volatility, while also maintaining a higher risk tolerance relative to traditional investors.

The Cabinet Office puts forward one of the most conservative estimates of the size of the OTC market, that is, more than 30 per cent of the total open contracts of the NYMEX market – enough to cause a significant distortion in the behaviour of capital in the market. In this vein, the report concludes that

> the volume of financial investments from commodity index traders and other investors with little or no specialist knowledge of oil markets may have allowed prices to rise beyond what would have otherwise been the case. (2008).

The flow of capital into the oil commodity market is therefore understood to have brought about a shot-term increase in spot prices, which in turn created positive futures expectations and higher prices. This behaviour came to an end when the financial crisis hit the financial institutions and all this capital was pulled off the oil market.

On the other hand, the OPEC puts forward one of the most exaggerated estimates of the size of the OTC markets in their annual World Oil Outlook reports, stating that

> speculator activity on the NYMEX surged to record highs in the first quarter of 2011. Open interest in the NYMEX WTI exceeded the unprecedented level of 1.5 million contracts which is 18 times higher than the amount of daily traded physical oil. (OPEC 2011)

In this statement, OPEC puts the size of the paper market into perspective, and emphasises the power of the financial investors in the oil market. In doing so, they propose that the OPEC is no longer to blame for fluctuations in the price level of the oil market.

In this blame-shifting operation, the OPEC attempts to shake off their reputation as eternal controllers of the oil price level and point to the responsibility of the financial actors in doing so. On the other hand, the UK Cabinet Office can be viewed as seeking to stress the necessity of regulating such wild markets, while trying hard not to upset one of the core industries that generates growth in the UK economy. Similarly, as the OTC market cannot be measured in any way, each institution or individual provides its own assessment based on its interests. The constant element, here, is that, however conservative or extreme the estimate, the size of this market is substantial enough to be able to influence the performance of the physical commodity markets.

When studying the data of the NYMEX oil futures for this period, the rise of speculative activity becomes evident. The total volume of oil futures increased exponentially from 2003 to 2007, while the increase in the volume of the total short positions is far greater than that of the long positions. The total commercial positions in NYMEX experiences a steady growth in the volume of both short and long positions, with a small acceleration after 2006.

On the other hand, in contrast to commercial positions, non-commercial positions experience an exponential increase in their volumes from 2003 onwards, falling in line with the first low-spare-capacity investments.

More specifically, what becomes evident is that the gradual increase in the volume of short and long positions during this period does not necessarily imply any radical increase in speculative activity. However, spread positions do experience a radical increase. The volume of spread positions, which is a speculatively driven investment (CFTC), grows from negligible in 2003 to almost the size of short and long positions combined in 2007/2008.

This argument is also supported by the data on oil futures prices relative to their maturity, in particular the breakdown among the one-to-five-year futures contracts. The price difference between the different maturities goes up at the end of 2004, only to fall down in mid-2008, when the oil market crashed. Indeed, this supports the argument that the oil market expected increased returns and that speculative opportunities were both widely available and highly demanded during this period.

The ratio of the commercial to non-commercial positions in the oil market during this period changed dramatically. In 2001, the ratio starts at about 6:1; in 2003, it rises at 4.5:1; in 2005, it suddenly leaps to 3:1; and, in 2008, it stops at a ratio of 2:1. According to these ratios, in 2001 there were as many as six commercial positions for every non-commercial one, while in 2008 there were only two commercial positions for every non-commercial one.

Hamilton (2009) suggested the possibility that positive-feedback speculation might have taken place in the oil market in this period. In these studies, 'positive speculation' refers to concept first put forward by Delong et al. (1990), whereby speculators buy commodity futures in response to rising prices, even though the underlying fundamental values are disconnected from those prices to the best of their knowledge. Hamilton considers the main determinants of the crude oil prices and finds that, due to the low price elasticity of demand, the increasing demand from Asia and the inability of production to adjust to it, these prices triggered the initial inflationary pressures on crude oil prices, and speculation followed. However, he stresses that the trigger point was derived from the underlying commodity market.

Another factor directly linked to financial speculation, which was considered to have influenced the oil price level, is the US dollar exchange rate. During the period after 2005, the US dollar exchange rate weakened significantly relative to its international counterparts.

A depreciation of the US dollar leads to an increase in the dollar value of oil; if it remains constant, the foreign currency cost of oil will be reduced, thereby boosting the international demand levels. Therefore, part of the increase in the price level of the oil market during this period can be attributed to the performance of the US dollar in the highly speculative exchange rate market. Nevertheless, in 2008, when the oil prices began to spike, the exchange rate of the US dollar was stabilised, highlighting the abnormality of the oil price spike of 2008 as well as the existence of speculative forces within the oil market.

According to Daniel Yergin (2008b), recent years have seen investors flowing from traditional financial investments to commodities. He observes that investors are seeking for stable investments in an era of uncertainty and turbulence, and the commodities-based financial investments provide them with just that. One may suggest that commodities are easier to understand compared to the more complex financial products: their price drivers are usually widely public and, although they are based on actual primary commodities with inelastic markets, they are not yet considered fully financial products. The oil market provides such an alternative investment possibility and therefore, especially in view of its performance compared to gold, attracted a vast amount of investments (Khan, 2009). The commodities markets now provided a very accessible market with positive returns, which was also able to hedge part of the inflation risk associated with many investment portfolios.

9.3 Independent Variables Shaping the Market Dynamics

Based on the behaviour of the futures market before the Iraq invasion, the oil price level was expected to remain at $20–$30 per barrel. This figure was the general expectation of the futures market, upon which trading took place. This expectation was formed under the assumption that nothing out of the ordinary would ever occur and that the analysts of the financial institutions, hedge funds, investment banks and index funds, as well as those of the oil-trading and oil-producing companies, were fully aware of all the long-term market dynamics, such as the rising Asian demand. Their expectations of oil price levels were shaped accordingly.

However, a number of factors made the investors reconsider their market expectations, in particular the war in Iraq, hurricane Katrina and labour strikes in Venezuela. The Iraq invasion is probably the main factor that reshaped the investor outlook as the other events were not close enough to the market or large enough to affect the market players' expectations, as well as market price levels in a substantial way. In the case of a demand shock, the necessary adjustments would have taken place within a 12-month period in order for market to rebalance and the oil price to normalise. This would also have been the case if the shock were more closely connected to hurricane Katrina. Yet, in this case we are talking about an almost continuous price increase from 2003 to 2008, when it ended with an abrupt drop at the end of the crisis.

Considering the links of this oil shock with the financial crisis that followed and the general environment of this time, many shared factors can be identified as influencing the developments in the oil and financial markets. The path of the oil price level is completely in line with the pattern of spare-capacity levels announced by the OPEC. Other patterns become more visible when considering the development of the housing bubble that took place in this same period.

The housing bubble had its roots in the reduction of the interest rates by the Federal Reserve in 2001, which resulted in the flooding of the markets with cheap money. The bubble kept inflating up to the mid-2006 period when it started decelerating until its subsequent fall in 2007. In other words, the booming period in the housing market, overlapping the increase in the oil price level, stopped prior to the boom period of 2007/2008. The commercial and non-commercial position data illustrate that speculation in the oil market started increasing from the mid-2003 and then experienced a sharp increase in early 2007 and late 2008, when the announced levels of spare capacity dropped significantly, spreading in its wake both fears of a peak oil and expectations of an increase in the oil price level.

9.3.1 *Drivers behind Speculation on Oil Markets*

This raises the question of what drove this rally in speculation levels, and consequently the spot price level in the oil market, and how this relates to the general financial and economic environment of the period. The first part of the answer can be found in the military activities of the USA in the Middle East. These activities had weakened the budget of the USA and had placed pressures on the US dollar value. In consequence, rampant oil prices had an increasingly negative effect on the macroeconomic performance of the USA, which now was bound to deepen its deficit by purchasing more volumes of expensive oil in a weaker currency.

It is commonly accepted that the international trend of increasing oil demand, disturbances in the oil supply and natural disasters accounted to some degree for the increase in the oil price level. Speculation remains, however, the key driving factor of this oil shock as well as the cause of its emergence. As discussed in the previous chapters, there are three factors that set the ground of a financial crisis: human behaviour, fundamental economic performance and technology and regulations. In the case of the 2007/2008 crisis, the levels of regulatory interventions in the oil market had been minimal and the CFTC had plans afoot to almost eliminate it by deregulating the approved commodity-based index funds (Clapp and Helleiren 2010). Thus, allowing for increased capital investments in oil market products.

As regards the fundamental economic performance of the market, the USA was not in a strictly dire position as, deficit and the exchange rate issues notwithstanding, it was otherwise relatively stable. At the beginning of 2007, however, everything changed. The unemployment level raised abruptly, GDP levels dropped and inflationary pressures started to spread through the economy. Studying the behavioural side of the period, the US consumers did not start losing their confidence in the economic and financial performance of their country before the end of 2007. There was, therefore, a lag between the turn in the economy and the public perception.

The same lag can also be observed in the performance of the US stock market, with the SPX index setting on a growing path starting from 2003 and till the end of 2007. This growing path is very much in line with the boom in the speculative activity in the oil market, but it seems disconnected from the pattern of the more financial real estate market, which starts contracting more than a year earlier. The growing pattern is explained by the consumer confidence index, however, as they start contracting simultaneously at the end of 2007. This challenges mainstream accounts of the real estate bubble, or subprime crisis, as the direct and only explanation of the 2007/2008 crisis. Although the connection is undisputed, the lag between the beginnings of the contraction of the real estate market and that of the stock market is evidence that the oil market also had a role in the development of this crisis.

Attempting to find the causality between these events, while also explaining the sudden change of the economic situation in 2007, raises a number of issues. First, the expectations of the speculative institutions involved in the oil market were that the price level of oil would keep rising and that, in doing so, such expectations attracted massive flows of capital into the market, thus creating a self-fulfilling prophecy. The conflicts in the Middle East, supply disturbances and increasing oil demand almost guaranteed that the positive performance of oil as a financial investment would be maintained; the second drop in the levels of spare capacity announced in 2007 further ensured that any disturbance would lead to even further increases. The attitude of the US government in its war against terror, its unwillingness to withdraw troops from the Middle East and its consequent surge of troops in Iraq to maintain order and stability, which started in mid-2006 and peaked at the beginning of 2007, also signalled that the US military involvement in the region would not end soon.

The developments described above, along with the steady and guaranteed increased returns that investments in the oil market seemed to present, attracted the inflow of vast amounts of capital. This change in investor preferences seems to have been in line with the peak of the housing market bubble, which began its deceleration as investors turned to the more short term and inflated profits of the oil market.

The significance of this claim is intensified by the fact that, while the housing bubble burst and the real estate prices kept falling in the following years, the oil market maintained its upward rally for more than a year before dropping back to its 2003 levels. It is important to point out that since 2007, when the price level of oil began rising, the effects of this hike had a direct effect on the macroeconomic performance of the USA. With the US exchange rate worsening and the price of its imports – including oil – fast increasing, the external aspect of the US economy was facing increased pressures. These pressures were promptly passed on to the levels of output and inflation, and from there to energy costs and unemployment. These dynamics put strong contradictory pressures on the economy from end of 2007 and led to the announcement of US official entering recession at the end of 2008.

To conclude, whether the investors chose the oil market as a rationally pondered investment alternative to the housing market or whether they were fleeing from a crashing housing market is not known. However, a linear causality between these events is apparent. This case clearly demonstrates the advancement of the financialisation process in the oil markets, which reached the level whereby the capital from other segments of the market, could flow freely into that of oil. By then, both the actors and the modus operandi of the financial element of the oil market would reach its maturity and become a rightful part of the international financial markets system.

References

Akerlof, G., & Shiller, R. (2009). *Animal spirits*. New York: Princeton University Press.
Baldwin, R. (2009). Integration of the North American economy and new-paradigm globalisation. *CEPR Discussion Paper*.
Clapp, J., & Helleiren, E. (2010). Troubled futures? The global food crisis and the politics of agricultural derivatives regulation. *Review of International Political Economy*, 19(2), 181–207.
Delong, B., Shleifer, A., Summers, L., & Waldmann, R. (1990). Noise trader risk in financial markets. *Journal of Political Economy*, 98(4), 703–738.
Freund, C. (2009). The trade response to global downturns: Historical evidence. *The World Bank Policy Research Working Paper*.
Hamilton, J. (2009). Causes and consequences of the oil shock of 2007–2008. *Brookings Papers on Economic Activity*.
Hardouvelis, G., & Stamatiou, T. (2009). Hedge funds and the US real estate bubble: Evidence from NYSE real estate firms. University of Piraeus.
Khan, M. (2009). The 2008 oil price 'Bubble'. Peter G. Peterson Institute for International Economics.
Masters, M., & White, A. (2008). The accidental Hunt Brothers: How institutional investors are driving up food and energy prices. Masters Capital Management and White Knight Research and Trading.
OPEC. (2011). *World oil outlook*. Vienna: Organization of Petroleum Exporting Countries.
Organization for Economic Co-operation and Development. (2010). *Working Party on Agricultural Policies and Markets Speculation and Financial Fund Activity: Draft Report Annex I*. Organization for Economic Co-operation and Development.
Portes, R., Dewatripont, M., & Freixas, X. (2009). Macroeconomic stability and financial regulation. *CEPR*, 178.
Stiglitz, J. (2008). Turn left for sustainable growth. *Economists' Voice*, 5(4), 1–3.
Tokic, D. (2011). Rational destabilizing speculation, positive feedback trading, and the oil bubble of 2008. *Energy Policy*, 45, 6009–6015.
Yergin, D. (2008a). Oil at the breaking point. *Testimony before the US Congress Joint Economic Committee*. Washington: US Congress.
Yergin, D. (2008b). *The prize: The epic quest for oil, money & power*. New York: Free Press.

PART III

Financialisation of Oil Market and Evolution of Oil Market's Actor Structure

# Financialisation of the Oil Market: The Four-Actor Structure

Abstract Against the background of analysis of the three phases of oil market's financialisation, this chapter proceeds with unfolding the actors' structure within the oil market and its evolution. In doing so, it answers the questions of how market actors evolved overtime and to which extent did the evolution of the market's actor structure influence the functioning of the oil market; the degree at which the new market structure dictate the relationship between the physical and the financial dimensions of the oil market and whether the understanding of the behaviour of the oil market actors can help one interpret market dynamics and oil price volatility in particular. In doing so, it provides evidence to the existence of the financialisation process being active within the structure of the oil market, as well as its transformative effects.

Keywords Financialisation · Oil market · Market actors · Volatility

Against the background of analysis of the three phases of oil market's financialisation, this chapter will proceed with unfolding the actors' structure within the oil market and its evolution. In what follows, the study will make a number of deep dives into the following questions: (a) how did the constellation of market actors evolve overtime and to which extent did the evolution of the market's actor structure influence the functioning of the oil market; (b) how does the new market structure dictate the relationship between the physical and the financial dimensions of the oil market and – last but not least – (c) how the understanding of the behaviour of the oil market actors can help one interpret market dynamics and oil price volatility in particular.

As discussed in the previous chapter, the element of speculation and price manipulation that evolved within the structure of the market through the effect of financialisation, as well as the gradual involvement of new actors in the market, strongly reinforced price volatility. Understanding of the role of price volatility for the new actors in the oil markets is arguably crucial for interpreting and forecasting their behaviour and, hence, impact on the market itself.

Now that oil is available for purchase as a financial asset, without ever taking position of the underlying product, speculation has been made not just possible, but also attractive, as argued by Khan (2009), very real. According to Frankel and Rose (2009), increased prices have amplified financial trading and, in turn, speculative activity in the oil market. The risk raised by the emergence of this form of speculation in the oil market is based on speculators' attempts at profitmaking through the manipulation of basic market fundamentals. This means that speculators can either sell a substantial amount of investments in the market, thus bringing its price down in order to repurchase it in a lower price, or inject a vast amount of capital, thus causing a rise in the price level and creating a bubble with a sudden inflow of investors.

As a result, at least up to the early 1970s, the structure of the oil market was shaped by three very important and influential groups of actors, whose power varied through time. The changes in the balance of power that have taken place among these three actors throughout the years have allowed for substantial changes in the functions of the oil market. First, the price of oil changed dramatically during the early 1970s, when OPEC took control of the market and precipitated the 1973 oil crisis. The period after the early 1970s has been characterised by a shift of control of the oil market to the supply side, with oil-producing countries dictating the supply levels of an increasingly inelastic market.

However, a new phenomenon emerged in the 1990s that introduced a fourth actor into the market; this was the start of the early financialisation phase of the oil market. The introduction of oil as a commodity that could be traded in the international financial markets marked the beginning of the oil market financialisation process, which introduced financial investment and, later on, speculation as integral parts of the market dynamics. The oil price level was no longer dictated by the three traditional market forces, since the financial aspect of the market had increasingly gained a dominant foothold in the workings of the market. This influence was reinforced by the rise of the new group of actors – or the new two groups of actors, if financial actors are to be divided into hedging and speculative actors on account of their distinct interests and modus operandi.

This conclusion raises the question of how the process of financialisation came to change these dynamics and what role the new group of actors play in the market. The first observation that must be made at this point is that the actors operating in the oil market have increased exponentially in number. In the past, the number of producers, importers and oil corporations had been more or less fixed. The new technologies and the process of financialisation, however, have now opened the market to almost anyone with a computer and an Internet connection. Even though the largest financial funds that have entered the oil market belong to financial institutions, the amount of individual investments should not be dismissed, as it is the combination of the two that is most indicative of the far-reaching effects that the process of financialisation has had on the oil market. As a result of this, a crash or a shock in the oil market today looks nothing like what it was forty years ago.

In the 1970s, when oil corporations only enjoyed meagre profit margins and oil-producing economies suffered a fall in profits, the effects were contained within the triangular structure of the market. On the other hand, the same crash today would not only affect these actors, but also cause massive capital losses to the international financial institutions or individual actors involved in the market. As these institutions and markets are international and unregulated, the effects of this crash spread much wider. Unrestricted by physical or financial barriers, the crash would go beyond the limits of the market structure and thus reach a larger percentage of world economies.

A second way in which the dynamics of the oil market have changed since the inception of the financialisation process is that the price level of oil is no longer determined solely by the interaction of the traditional market forces. The oil futures market performance is directly linked to that of spot prices, which entails that the new actors in the market structure have a direct influence over the price of oil. In the absence of speculation, the main interest of financial investors is to increase the value of their investments. Hence, as long as the price level of oil rises, more financial investments will enter the market and push prices further up. In this view, the new dynamic of the price-setting mechanism is one where one group of actors are bent on rising oil prices.

Nevertheless, not all financial actors are interested in raising the price level of oil, as speculative positions on index trading or in spreads can be taken in the possibility of a change, upwards or downwards, of the price level. As a result, this new group of actors has a primary interest, that is, increased price levels and a secondary motive, increased market volatility. This group of actors, however, does not have the power to influence the market in a direct way, as producers do when restricting output levels. The only exception is during extensive and targeted speculation when the price level can be manipulated to a significant extent, as discussed above. As a result of these two developments in market dynamics, the effects of volatility are now being felt by an increasingly wider spectrum of actors in an increasingly wider geographical and demographic area.

There is evidence to the effect that the oil market has gone through a process of evolution since its deregulation and financial marketisation and that, in particular, the volatility of its price levels has seen abrupt and extensive upward changes. There is evidence, also, to show that the oil market has undergone a process of financialisation with the emergence of a new group of actors who has built links to the asset markets and, with the help of an emerging class of speculators in their ranks, has altered the dynamics of the traditional price-setting mechanism of the market.

10.1 The New Group of Actors on the Oil Market: Characteristics and Behaviour

The analysis of the three phases of oil market's financialisation demonstrated that the petroleum futures market expanded both in size and in number of participants. With privileged access to data from the CFTC, Büyükşahin et al. (2008) were able to confirm this trend. Indeed, this evolution of the financial actors' structure, both in the underlying commodity and the oil-based financial markets, entails an increase not only in the ranks of actors, but also in their variety. When it comes to actors active in physical oil markets, a number of large International Oil Companies (IOCs), large Utility Companies, commodity trading companies and small oil-producing and storage firms operate in asset markets mainly as a means of diversifying their portfolios or hedging their risks.

National Oil Companies (NOCs) are seldom involved in these markets as they operate under government control. In the case of financial institutions' involvement in asset markets, on the other hand, the main financial actors include hedge funds, investment banks, asset management institutions and specialist trading houses. Generally speaking, there are two major categories of purely financial investors: active managers and passive index managers. The first category consists mainly of hedge funds and pension funds actively engaged in the oil-based product market, while the second is more recent as it first appeared at the turn of the millennium when market products and indexes had already become widely available and easily accessible.

In recent years, the investment activities of such institutions have intensified not only in volume, but also in character. This is because some of the largest financial institutions have started investing directly into the physical production of the oil commodity and its products. JP Morgan Chase and Goldman Sachs, for example, are two of the largest crude oil suppliers, and buyers of products, of the refineries owned by Alon USA, California, Louisiana and Texas, as well as of the Carlyle Group PBF Energy and other New Jersey and Delaware refineries (Meyer 2001). Large oil product consuming companies, such as energy suppliers and distributors, as well as airlines, are also getting increasingly involved in these markets by investing directly in refineries or storage facilities. This development goes against such academic findings, as that of Campbell and Viceira (2003), who noted that the financial investors in the oil market, for example, hedge funds, characteristically concentrate their investments on the most liquid parts of the oil futures market in order to retain the freedom to retrench as needed.

Nevertheless, the above-mentioned phenomenon can be explained by reference to two major factors. On the one hand, high oil prices have reduced the profit margins of the oil downstream sector; this has made investing in refineries cheaper. On the other hand, recently introduced regulations on commodity trading, such as the Dodd-Frank Act of 2010, have induced these institutions to find alternative ways to increase their exposure to this market. Along with these two main drivers, other factors also played a part, such as the security of supply and the commercial requalification of previously non-commercial institutions on the back of their successful involvement in the market.

As can be seen in Tables 10.1 and 10.2, drawn from Bassam Fattouh (2011), financial and oil market actors have partaken in both types of

Table 10.1 Participants in the 21-day BFOE market and their shares in trading volume (Fattouh 2011)

	Sales (b/d)				Purchases (b/d)			
	2007	2008	2009	2010	2007	2008	2009	2010
Arcadia	–	–	–	–	485	–	–	–
BP	23,786	3,005	13,699	29,545	25,243	273	10,959	12,662
Chevron	–	273	274	–	–	273	–	–
ConocoPhillips	18,447	11,749	12,329	32,143	6,311	5,464	12,329	29,545
Glencore	–	–	274	–	–	546	548	–
Hess	–	–	9,315	37,338	–	–	10,137	20,779
Hetco	–	–	822	7,143	–	–	1,096	974
Mercuria	12,136	12,842	64,658	79,545	13,107	24,863	54,247	89,286
Morgan Stanley	–	–	274	28,896	–	–	3,014	19,805
Noble	–	–	548	6,494	–	–	822	5,844
Philbro	46,602	19,126	25,479	23,377	36,408	23,770	36,164	14,935
Sempra	15,534	18,306	13,151	8,766	18,447	19,672	13,699	7,792
Shell	34,951	62,022	125,205	91,883	46,117	32,787	73,151	75,000
StatoilHydro	–	273	–	–	–	–	–	–
Total	–	–	–	649	–	–	–	2,273
Totsa	31,068	16,667	53,425	62,987	61,650	28,962	108,767	83,442
Trafigura	–	–	–	16,234	–	273	–	10,714
Unknown	–	273	–	–	–	–	–	–
Vitol	68,447	12,842	48,219	56,818	43,204	20,492	42,740	108,766
	250,971	157,378	367,672	481,818	250,972	157,375	367,673	481,817

Table 10.2 Participants in the crude oil futures market (Fattouh 2011)

	Sales (b/d)				Purchases (b/d)			
	2007	2008	2009	2010	2007	2008	2009	2010
Addax	–	–	411	–	–	–	740	812
Arcadia	23,301	4,918	4,658	14,448	6,553	10,109	6,575	17,208
Astra	–	–	–	–	2,427	1,298	–	–
BNP Paribas	–	–	548	5,519	–	–	2,192	4,221
BP	26,214	55,601	74,085	76,948	43,083	37,432	24,397	75,010
Cargill	485	1,913	411	–	485	4,918	274	1,136
Chevron	17,233	26,093	70,699	84,659	43,811	47,541	53,863	73,195
Chinaoil	–	–	–	–	–	–	1,233	–
ConocoPhillips	485	10,410	23,041	33,766	728	24,863	28,630	60,065
Glencore	1,456	1,940	14,219	24,968	485	4,372	9,863	26,299
Guvnor	–	7,240	13,151	3,571	1,942	5,464	3,836	1,299
Hess	971	273	2,192	22,240	–	–	2,192	17,532
Hetco	–	–	–	3,571	–	–	1,092	974
IPC	–	273	2,055	325	–	1,481	3,068	1,786
Iplom	–	–	548	–	–	1,093	548	1,136
Itochu	–	546	7,671	7,253	–	6,126	11,041	10,844
JP Morgan	9,223	11,380	9,153	7,792	1,456	29,358	54,973	14,935
Koch	33,010	36,284	23,556	3,247	11,165	37,205	34,849	32,305
Lukoil	971	13,798	28,559	24,513	485	7,049	20,411	21,753
Maesfield	–	–	1,644	1,136	–	–	–	1,623
Marathon Oil	–	–	–	–	11,408	9,699	548	6,494

(continued)

116 SPIKES AND SHOCKS

Table 10.2 (continued)

	Sales (b/d)				Purchases (b/d)			
	2007	2008	2009	2010	2007	2008	2009	2010
Mansfield	–	273	1,233	3,247	–	3,825	685	–
Mercuria	34,345	46,809	59,726	79,471	31,311	68,415	99,841	117,156
Merryll Lynch	1,942	4,781	1,918	1,299	7,646	4,645	–	–
Mitsubishi	–	–	–	–	–	–	3,014	–
Mitsui	–	273	–	–	1,456	546	–	–
Morgan Stanley	20,338	24,317	57,882	100,487	20,146	17,760	51,238	88,377
Murphy	–	–	–	–	–	410	–	–
Natixis	–	–	42,033	19,968	–	–	36,849	27,110
Neste	971	4,372	2,740	–	–	3,005	822	1,623
Nexen	1,942	4,577	4,003	6,951	2,427	2,691	5,189	11,685
Noble	–	–	822	14,286	–	–	548	8,442
OMV	1,485	–	14,562	28,545	–	5,787	36,995	48,880
ORL	–	1,093	–	–	–	2,186	–	–
Pe traco	–	820	1,644	974	–	1,735	2,192	–
Pe trodiamond	–	–	–	1,948	–	–	822	–
Pe troplus	5,583	3,825	1,918	–	1,942	–	1,644	–
Philbro	20,146	48,656	68,923	82,867	36,772	52,117	34,400	50,487
Pioneer	–	–	–	–	–	–	137	–
Plains	–	2,732	–	–	–	–	2,466	1,299
Preem	–	–	685	–	–	–	3,562	–
Sempra	971	7,978	9,644	2,273	4,854	15,929	15,616	2,922
Shell	47,694	131,929	132,079	149,221	52,699	39,727	83,995	129,545

Sinochem	—	—	—	1,136	—	974
Sinopec	—	—	1,932	2,597	—	1,867
Socar	—	—	—	25,000	273	9,091
Sonatrach	—	—	274	974	—	8,279
Standard Bank	—	—	932	5,575	—	5,195
Statoil	6,796	2,186	8,630	108,224	—	118,130
StatoilHydro	14,563	77,945	59,233	—	273	325
Totsa	19,782	23,087	45,260	25,974	54,781	60,575
Trafigura	971	16,940	29,315	27,955	46,325	32,649
Unipec	8,738	7,377	4,521	8,955	13,798	11,578
Valero	14,456	546	1,096	—	12,432	54,545
Veba	—	—	—	—	14,208	—
Vitol	36,044	58,579	132,060	245,692	1,093	98,214
	350,166	639,764	959,666	1,257,575	49,795	1,257,575
				337,165	639,764	
					959,661	
					112,447	
					19,726	
					29,170	
					28,877	
					47,397	
					61,863	
					8,630	
					548	
					7,260	
					—	
					2,800	
					603	

markets. Table 10.1 presents the list of the participants in the 21-day Brent FOE market, which is a market for trading forward contracts of 600,000 barrels of Brent oil. This market is used by oil companies either for trading their crude oil productions or for purchasing crude oils for processing. As illustrated in Table 10.1, the composition of the actors involved on the buying side of this market is dominated by a few large IOCs, as well as by a few large oil-trading companies; of particular note, here, is Morgan Stanley's steady involvement since 2009. As trading of oil commodities is mostly ducted through bilateral agreements between producers and consumers, the volumes traded in this market are those of small oil producers. Requiring forward contracts to finance their productions, however, small oil producers represent only a small fraction of the kind of underlying crude oil trading market that takes place on commodity trading platforms.

Table 10.2 contains a list of the major participants in crude oil futures market between 2007 and 2010. This table shows a wider range of participants. On the one hand, JP Morgan, Morgan Stanley, Standard Bank, BNP Paribas and Merrill Lynch are some of the dominant international financial institutions in crude oil futures trading. On the other hand, the IOCs are leading participants in the oil market, either in the upstream or downstream side, together with some of the largest oil commodity trading companies, which also trade large volumes of contracts. That said, two important caveats must be noted with regards to the data in this table. First, the volumes of trades reported are of Central Counter Party (CCP) trades and do not include any OTC trades in which these companies might be involved. Second, the absence of NOCs from these trade figures confirms the extent to which their trade focus – mandated by central governments rather than market incentives – remains on the underlying commodity market.

Figure 10.1 adds to the findings of Tables 10.1 and 10.2 with a breakdown of the major participant groupings in the futures oil market. These figures come from the UK Cabinet Office and are based on CFTC data, where market participants are divided between commercial and non-commercial. The CFTC collects data from CCP trades, where traders are obligated to declare the purpose of each trade order as either commercial and hedging or non-commercial and speculative. The CFTC then makes these data publicly available. According to these data, non-commercial positions are dominated by hedge funds that account for the largest percentage since 2006. At the same time, the degree of participation of floor brokers and traders representing

10 FINANCIALISATION OF THE OIL MARKET: THE FOUR-ACTOR STRUCTURE 119

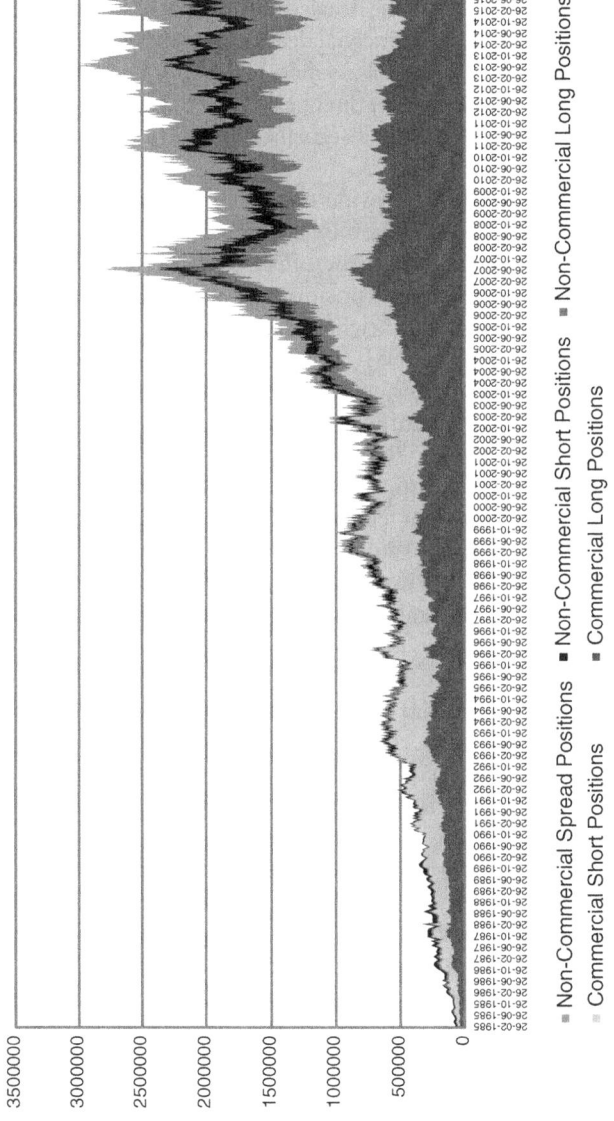

Fig. 10.1 Composition of the recorded oil futures NYMEX (Bloomberg Data 2015)

smaller institutions or individual investors can be also considered quite remarkable. On the commercial side, manufacturers and producers of oil, account for only a fraction of the total trades in the market. At the same time, commodity swaps and commercial dealers take up the widest percentage of this market. However, these dealers often act as intermediaries for financial institutions who do not want to get direct exposure to oil market trading (UK Cabinet Office 2008); thus, their classification as commercial trades is questionable.

Masters and White (2009) provide evidence that proves that by March 2008 the oil futures index had risen by a quarter of a trillion dollars. They also put forward an argument to the effect that the distinction between commercial and non-commercial positions as published by the CFTC is wrong. According to the CFTC, when trade takes place within a clearing house, such as the NYMEX or the ICE of London, the trader has to mark his order, on a trade-by-trade basis, as either commercial or non-commercial, depending on the type of institution that requested it.

This raises questions over the motives of each institution, as well as the idea that some institutions occasionally act as intermediaries, thereby diluting the ratio of commercial and non-commercial trades (UK Cabinet Office 2008). While the US CFTC estimates that about 20 per cent of the NYMEX oil futures trades are non-commercial, Masters and White (2008) calculate that the actual ratio of commercial to non-commercial futures is significantly higher. Nevertheless, the UK Cabinet Office report claims that the difference in investment behaviour between financial and traditional investors is traceable, and indeed comports with the known evolutionary track of market behaviour. It is claimed that traditional financial investors use oil as a hedging mechanism for their market investments (commercial positions) and that, therefore, they generally hold long positions and are not easily affected by short-term developments in the market.

Figure 10.1, in addition, shows a steady, and very conspicuous, increase in the number of both commercial and non-commercial actors in the market, especially since 2000. However, the percentage increase of non-commercial positions in this period far exceeds that of commercial ones. The same trend was also announced by CDA/Wiesenberger, which reported that mutual funds had grown in number and assets, respectively, from 155 and $16 billion in 1959 to 10,725 and $6.9 trillion by the year 2000.

10.1.1 Speculative Behaviour of New Market Actors

Looking more closely into the composition of non-commercial trades, Fig. 6.1 plots them with respect to their long, short or spread positions in the oil market. A number of conclusions can be reached from this. First, the number of non-commercial contracts has risen dramatically since the end of 2000. Second, all three of the available types of contract have increased in volume since then. Third, the spread contracts, which represent mostly speculative investments, recorded the highest increase, while long positions, which represent mostly hedging or traditional investments, come a close second. Even short positions, the least popular, have in fact more than tripled in volume since 2000. These figures suggest that the developments of 2000 led to a massive inflow of capital in the market, which then spread across all the different products available. What is more, they suggest that, even though all three types of product increased over this period, the more speculative ones enjoyed the greatest increase, which heralded the emergence of speculation-driven investment interests.

In keeping with these findings, the CFTC (2008) considers speculation in the oil futures market such an important aspect to the oil price-setting mechanism that has begun holding hearings on the possible limits on futures speculation in the energy market. At the same time, the UK Prime Minister Gordon Brown and the French President Nicolas Sarkozy penned a joint article in the Wall Street Journal (8 July 2009) about the damaging effects of speculation on the global economy, calling on the International Organisation of Securities Regulators to regulate the oil futures market and to examine its influence on the system of oil pricing. Interestingly, a similar call for an investigation of the effects of oil futures speculation on oil prices was also issued by the US Senate Permanent Subcommittee in 2006 – years before the crash.

According to the UN Trade and Development Report (2009), the inflow of new investors into the oil market has multiplied instances of herd behaviour and consequently intensified the decoupling of the financial aspect of commodity markets from the dimension of real production, especially in the short term, when the demand and supply forces of the commodity markets are particularly inelastic. This is because short-term investment shocks in the financial aspect of commodity markets have the ability to alter the price level of these oil-based financial products. On the other hand, the inelasticity of the supply and demand of the actual oil commodity can produce a mismatch between these two markets and the creation of speculative bubbles.

Tang and Xiong (2009), in their study of the functions and effects of the financial aspect of the oil market, have traced an increase in the influence of world equity shocks as well as US dollar exchange rates shocks on commodity index rates, especially in the post-2000 era. They conclude that the financialisation process has tangible effects on the commodity markets. By way of example, they compare the stronger responses received by the two biggest commodity indexes, the GSCI and the DJ-AIG, to the performance of similar, but un-indexed, commodities and, in doing so, they bring to light the spillover effects of volatility from the financial markets to the commodities markets (Silvennoinen and Terosvirta 2009).

This process of integration between commodity markets and asset markets can be understood by reference to different factors. First among them is the fact that a growing number of investors have begun to hold a combination of both commodities and financial assets in their portfolios. The herding effect, however, drives them to follow very similar hedging policies and to invest in very similar products; this makes them mediators of financial shocks between the two increasingly mutually vulnerable markets. Due to portfolio trading, for example, poorer performance in one market causes liquidations in the other markets in the portfolio as their traders attempt to readjust their positions (Kyle and Xiong 2001).

Harry Markowitz's portfolio theory (1952) suggests that since stock price movements are random, it is not possible for an investor to predict how to beat the market. Thus, in order to optimise their returns against risk, financial investors have to hold investments that equate to the entire market. The high risks of this sort of investment can be bypassed through a portfolio diversification strategy where the capacity to quantify the risk–return trade-off of financial assets, along with their correlated returns shifts the risks to the centre of investment decision-making. Thus, investors evaluate assets according to their variance and co-variance of returns over time. As a result, volatility becomes a crucial determinant of the investment process insofar as returns are counterbalanced by risk tolerance (Wigan 2008).

Furthermore, increased co-movement is observed among commodities that are only related by virtue of being part of the same commodity index and being traded in that capacity. This co-movement implies the financialisation of these commodities in correlation with the asset markets (Pindyck and Rotemberg 1990).

The increasing number of new participants in commodity markets can be typically characterised as holding large positions and being as prone to irrational investment behaviour. Their activities have intensified the decoupling of commodity futures price levels from their fundamental value. In other words, the increasing involvement of financial institutions in commodity markets, as well as the novel ways of marketing its products, has coincided with the financialisation of these markets, insofar as they behave increasingly like asset markets.

Labban (2010) offers a performative economic theory take on this particular development. In his view, the emergence of oil as an asset market has created a financial market parallel that is uniquely integrated with its underlying commodity market. He argues that big banks and financial institutions, which had taken up high risks when first entering the underlying oil commodity market, have become themselves market makers as they started taking positions of their own in the market. Most prominently, they employed oil derivatives, originally designed to protect market actors against volatility, as part of the development and construction of a new asset class, the commodity index, that generated profits independently from the actual production of oil and thereby allowed for the rise of speculation and further volatility.

According to the definition of financialisation furnished at the outset, financialisation occurs when the emergence of new financial markets creates new processes and alters existing financial investment interests, within and outside the market structure, thereby attracting further new financial actors and institutions. The oil market seems to fit the pattern. The emergence of a new type of market via the oil futures and options market has allowed oil to be traded – and increasingly so – as a financial asset, independently from the underlying commodity traded as contracts with physical delivery. The very nature of this new market has attracted a vast amount of capital deriving from purely financial actors and institutions, whose motives are financial and unrelated to the underlying product linked to their trade assets.

In line with the definition of financialisation given above, the number of these new institutions and actors involved in the oil-based financial market, along with the capital injected into it, has grown rapidly, especially since 2005. What is more, the development and widespread adoption of new technologies, as described above, contributed in no small part to this market growth. When oil futures and options became available to anyone with an Internet connection, in 1999, it can safely be argued that the processes and functions of this market changed irreversibly.

The process of financialisation in the oil market has brought about an unprecedented growth in the number of actors and volume of capital active in its structure. While prior to the 1980s and early 1990s the number of financial institutions and actors involved in the oil-based financial markets was restricted by regulation, accessibility and lack of product and trade information, the developments of the late 1990s and early 2000s made access to the market possible and attractive. Institutions that had no prior involvement or knowledge of this market were rushing to find a way in. Masters and White (2008) measured that the value of assets allocated in commodity passive index funds had risen from US$13 billion to 260 billion within a period of five years starting from 2003. As disparate sources as a Staff Report of the US Senate Permanent Subcommittee on Investigations (2006), Masters and White (2008) and later Parsons (2010) agreed that trading by speculative financial institutions in the oil market had pushed the futures prices to levels that could not be justified by rational expectations of future supply-and-demand patterns. Joachim von Braun was quoted as saying that "we have good analysis that speculation played a role in 2007 and 2008 [...] Speculation did matter and it did amplify, that debate can be put to rest" (Irwin et al. 2009).

Data from the study of Daniela Tavasci and Ventimiglia (2011) on the financialisation of primary commodities clearly indicate that, even in a small sample of financial institutions involved in index trading, there has been an abrupt increase in the volume of capital invested in commodity futures. Their research also maintains that according to the Bank for International Settlements (BIS), "the number of futures and options contracts outstanding on commodity exchanges has increased about fivefold and the notional amount of outstanding OTC commodity related contracts reached the astronomical figure of $13 trillion in 2008" (UNCTAD 2009). They conclude that commodity derivatives have transformed into a proper asset class and are regarded by financial investors and traders as just another financial asset in the composition of their portfolios.

Following from this, the triangular structure that had dominated the market dynamics up to the late 1970s underwent a radical transformation in the period between 1980 and 2010. The process of financialisation has opened the structure of the oil market to a new group of actors who are motivated by purely financial interests and bear no connection to the underlying commodity of the market. The spot price of oil, as a result, is no longer determined only by the interaction of producers, consumers and mediators. This is because a fourth actor is now responsible for the lion's share of the financial capital invested into the oil market and is therefore actively involved, and increasingly influential, in the price-setting mechanism.

10.2 NEW FINANCIAL MARKET PLAYERS TRANSFORMING THE OIL MARKET: OIL SPOT PRICES AND FUTURES PRICES IN FOCUS

The relationship between oil spot prices and oil futures prices has been a subject of contention in recent years. A point of agreement in the literature, however, is financial markets play the central role. As compellingly articulated in Kaufmann (2011), the particular evolution of the financial markets over the past decades has been responsible for introducing 'speculative' factors into the price mechanism. Simply put, after the addition of financial actors to the traditional triangular structure of the oil market, oil prices have progressively lost their correspondence to the fundamental value of the underlying market commodity.

Part of the reason for this is that the new financial actors follow different investment patters compared to traditional oil investors. Financial investors generally use oil investments as a hedging mechanism in their investment portfolios and, on account of their similar access to information, knowledge and expertise; their investment patterns are generally synchronised and easily identifiable. As they mostly hold long positions, which, in the context of the commodity futures and options markets, are indicative of an expectation that the value of the commodity will rise, they are not generally influenced by short-term fluctuations and market volatility. At the same time, they maintain a higher risk tolerance relative to traditional investors. Against this background, a UK Cabinet Office Report (2008) calculated that financial investments account for more than 30 per cent of the total open contracts of the NYMEX market and have extensively shaped the pricing of this market.

Labban (2010) proposes that the financial performance of the oil products influences the performance of its spot price despite the fact that they are traded in physically separate areas. This is because the performance of financial products may be acting as a guide for oil spot market traders. However, the financial oil product pricing is based on future expectations of the conditions in the underlying commodity market – hence the circular pattern of influence between the two entities. Labban also calls attention to the fact that financial markets frequently overreact to these expectations, especially compared to the underlying commodity markets, which, as a result of their mobility and liquidity, allow the speculative effects of financial overreactions to penetrate the core of their relationship. In a rather carefully formulated passage, Labban notes that

> The spot price is driven by speculation on conditions in spot markets mediated by the reaction of financial markets to such speculation [...] fundamentals [...] determine oil prices only by way of the effects on financial markets of speculation on future conditions in spot markets.

Primary research conducted for the purposes of this book shows that Platts, currently the dominant international data provider of physical commodity price reporting, proceeds from this very understanding of the impact of financial markets on the oil price-setting mechanism. Platts supplies real-time data and reporting to the majority of the international oil market actors, from the majors to independent oil producers, refiners, traders and even news agencies specialised in oil market reporting. Their data are considered the most accurate and reliable source in the market, and are thus employed for the drawing of reports or as a way to monitor market performance and to influence expectations and decision-making. Understanding how the data published by Platts are collected and published will be remarkably revealing of the relationship between the physical and financial aspects of the oil market.

To this end, the author of this book attended workshops especially organised and run by Platts to present what they call the Platts Methodology. These workshops simulate the publication of data for a specific grade of crude oil, but the same methodology applies across all of the products they cover. The process starts in the morning and ends with the determination of the value that the actors involved deem representative of the price level of the commodity on that day. The value then is added to the series of daily data that are used to draw the time series of crude oil prices.

The procedure begins with the publication of a value on the online platform that is available to all the actors subscribed to Platts. This is the value that Platts analysts believe to represent how much the given product is worth on that day. This value is obtained using a model that considers a range of inputs, such as the final value for the previous day, the expected value for the current day according the futures market of the previous day, as well as any recent unexpected developments affecting the performance of that particular market. Once this value is determined, it is published on the online platform, and quotes for 'buy' and 'sell' interests or for actual trades are then fed back to the system around this value.

Buy quotes are only accepted below the value of the initial price, while sell quotes only above it, up until the final stages of the price determination process when the barriers are lifted and quotes can be placed freely. This triggers a bargaining process that leads the quotes to converge towards the initial value up until the release of the restrictions to buy and sell quotes. However, high deviation from this value is rare in the closing bargaining stages in the absence of a market-shaping event, as the positions have already been set and the price level has been accepted among the actors.

The window for submitting quotes closes at 18:00 GMT every day, so market actors have a time up until the official close to submit their interests or trade agreement values. The value that is reached at the end, as the buy and sell values converge to the market value, is published as the market on close value and is considered as the fair value of the commodity in the market according to the market actors. If a single value between the buy and sell quotes has not been reached, then the average is published after approval by Platts analysts. This methodology is highly valued by market actors as they are directly involved in the price-setting process of the market in which they operate.

This process, however, places a strong emphasis on the links between the physical and financial aspects of the oil market. This is because the market on close value is considered the price of the given crude oil, or oil product, in the physical market, and it is the value that is then published by newspapers, reported in the media, and assumed in strategic decision-making the world over. However, this value is shaped directly by two different dynamics. The first is the market itself, which comes into play when price-level precedents are factored into the methodology and when market actors interact to produce the market on close value. The second dynamic is the futures market, as expectations of today's price level are formed on the basis of yesterday's futures prices.

This process shapes a performative cycle between the two aspects of the market, as the futures-based expectations of oil value, which is determined by the financial aspect of the oil market through the use of financial models, is used as a benchmark against which current physical trading takes place and market value is negotiated. This is not to say that the value set by the Platts Methodology model, which employs futures expectations, will always be equal to the market on close value – although, according to Platts, this is usually very close. In return, the value of the futures prices is determined as a function of the current price level of the physical market as well as the expectations of its future performance. As a result, what can be observed is that the determination of the physical and financial price levels in the oil market are interdependent, directly affecting each other in a mutual feedback loop.

Timo Behr (2009) claims that in case of market tightening in the structure of the underlying oil commodity market, the spot price of oil will increase. The financial markets will then adjust their expectations of the future price of oil under the new spot price, which translates to higher future prices as new capital enters the futures market in an attempt to capitalise on these positive expectations.

As oil futures prices increase, any producers in the underlying commodity market, or any actors with the resources to store oil reserves, will refrain from selling their stocks at current spot prices because they have the option of selling them at a later date for a higher price and, hence, higher returns. This behaviour is known as the storage arbitrage condition, and it ultimately leads to a further tightening of the oil spot price market, as less supply of commodity is made available in the market. On absence of an external event, the cycle only reinforces this pattern. In such cases, the body with the power to intervene on the market price is usually the OPEC, although, in some cases, economies with substantial oil reserves have the ability to release them in order to rebalance the market.

An episode of just this sort of performative cycle between the physical commodity market and its oil-based financial products occurred in September 2012, when the financial actors in the oil market reacted to the possibility of an increase in the supply levels from the Middle East by seeking to liquidate their positions under the expectation that the supply levels would not be absorbed by the markets and would eventually drive prices down. Sure enough, Christopher Bellew of Jefferies Bache, a commodities and derivatives brokering house, stated that "people are thinking that maybe the Saudis are going to produce more, and some funds are taking the opportunity to liquidate positions" (Choy et al. 2012).

Kaufmann and Ullman (2009) studied the links between the spot prices of crude oil and its futures prices. They propose that prices set by the crude oil futures eventually get passed on to the spot market. As financial actors, such as speculators and investors, purchase futures with the expectation of a price hike based on their assessment of the fundamentals of the market, the spot prices increase through their link with the futures. Therefore, they contend that any explanation of the oil price spike of 2008 must consider speculative activity as much as market fundamentals.

Nevertheless, other authors, such as Silvapulle and Moosa (1999), and more recently Bekiros and Diks (2008), have proposed that the links between the spot prices and the performance of the futures market are bidirectional and not linear. These studies show that both markets react to changing conditions in a synchronised manner, and that the pattern of leads and lags varies over time. Silvério (2010), while cautioning against overlooking the role of macroeconomic market fundamentals, seeks to find an empirical measure of the degree to which financial markets influence the price-setting mechanism that regulates the spot markets of crude oil benchmarks. Sure enough, evidence to this effect is found in relation to the futures market and the behaviour of speculators, institutional investors and commercial hedger. Their conclusions confirm the findings of previous studies conducted by Kaufmann (2011), Tokic (2011) and Cifarelli and Paladino (2010).

10.2.1 *Expectations and Behaviour of the Financial Actors*

Owing to the sheer variety of individual actors, a single universal factor, or combination of factors, is surely hard to come by. An attempt can be made, however, at identifying a limited number of broad causal factors. Any oil market investor, for example, will be influenced, among other factors, by the current oil price level, past performance patterns, global geopolitical events, fundamental macroeconomic indicators and the international stock market performance.

One factor in particular, though, takes pride of place among the lot: the spare-capacity indicator produced by oil-producing countries. It has been noted that this indicator is of questionable accuracy and often fallaciously linked to peak-oil fears. Nevertheless, as low spare-capacity levels imply the possibility of a supply shortage, its role in financial investment behaviour is unquestionable.

Analysis of the nature and behaviour of financial actors would be incomplete without reference to what drives their involvement in the oil market. Here, financial investors can be separated between those interested in diversifying their portfolios and those interested in pure speculation. With regards to the latter, despite the lack of reliable data on its exact level of pervasiveness, evidence exists that speculation, especially since 2001, has had an impact on the volatility of price levels in commodity markets (UNCTAD 2009). Even though empirical studies by Chong and Miffre (2010) and by Büyükşahin et al. (2008) conclude that the correlation between commodities and stock returns is not significant, other studies have had opposite results by focusing on the study of increasing prices as the main drivers of speculation.

In order to find a strong indicator of speculative activity in the oil futures market, analysts in oil firms have attempted to study the movements of spot prices by tracking the volume of net open positions, rather than total volume (Masters and White 2008). In doing so, they separate the commercial from the non-commercial positions and calculate the influence of speculation on the basis of the former, as non-commercial positions are believed to be used almost exclusively for the purposes of risk hedging. Their results demonstrate that the volume of net open positions increased by more than 200 per cent between 2003 and 2008, while non-commercial positions reached their peak during the same period (CFTC 2009), thus proving that speculative activity was marginal.

In an attempt to identify the levels of speculative activity in the oil market, Khan (2009) proposes a comparative analysis of the behaviour of the oil and gold prices as the gold market is already known as a commodity market dominated by speculation-driven investment interests. He hypothesises that the correlation between the performances of the two markets would provide compelling evidence of the kind of investment interests that permeate the oil market. Sure enough, his results confirm a very close relationship between the performances of the two commodities, which leads him to conclude that speculative activity has indeed been taking place in the oil market, especially in the post-2000 period, when the increase in the price of oil accelerates past that of gold. Khan proposes that this decoupling can only be explained by an uptick in speculative activity as it became more profitable to invest in oil than in gold. This argument becomes even more compelling when considering that the speculative wave came to a dramatic end after 2008 as the drop in the price of gold, by only 19 per cent, paled in comparison to the collapse of the price of oil, which had gone down by about 70 per cent.

In a similar vein, Tavasci and Ventimiglia (2011), in their study of the financialisation of the copper market during the 1994–2008 period, find a dramatic change in price level and an alignment of the copper futures and spot prices from 2003 to 2008. In their account, the three most cited explanations for this price pattern are: market fundamentals, in terms of increased demand from China and India (Gros 2008); speculation in the commodity markets (Wray 2009), where actors, usually in the form of oil majors, restrict their outputs in order to trigger price hikes; and finally, the 'parallel' commodity futures market. On this front, she contends that the deregulatory events of the commodity markets, which started in the 1930s and lasted until the 1990s (when the CFTC started granting Wall Street exemptions on the limits of positions and OTC swaps), contributed to a transformation of the copper market whereby index managers were allowed to take on positions. Especially after 2003, she observes a boom in price levels, which she explains as a result of the increased levels of international liquidity.

According to the two studies above, the oil market has become increasingly financialised, especially with the intensification of deregulation, technological advancement and financial marketisation since the early 2000s. This process has transformed the commodity market into an asset class, where financial actors increasingly dominate the dynamics of the oil market by injecting vast amounts of capital and, at the same time, creating direct links between the commodities and asset markets. Within the group of financial actors, studies have revealed the existence of a subcategory of speculative investors and price manipulators. These actors are especially active in the markets of short and spread products and usually follow – or, indeed, drive – noise-trade strategies in order to inflate these markets and take advantage of their fluctuations.

Therefore, the effects of the financialisation of the oil market are twofold. On the upside, the oil market has attracted vast amounts of capital and a larger cohort of market participants. On the downside, this financialisation process has also caused an increase in the spot price level of oil due to the introduction of speculation and price manipulation, which has upended the traditional price-setting dynamics of the market. Not surprisingly, higher prices are usually welcome by oil-producing countries, oil corporations (be they NOCs or majors), and even financial investors interested in increasing the price of their investments. This means that as long as the price of oil does not rise to levels that would encourage recourse to alternative energy sources, its negative impact in the market is sustainable.

In addition to this, the inverse alignment of the oil market behaviour to that of the asset markets after the turn of the millennium has produced increasing volatility in price levels (Pindyck and Rotemberg 1990). UK Cabinet Office (2008) produces data on crude oil spot price monthly volatility since the 1970s, which propose that volatility levels have increased noticeably since the late 1980s. Based on a monthly index of oil and other commodity prices covering the period between 1945 and 2005, Regnier (2007) proves that oil price volatility has increased significantly. Duffie et al. (2004) reach the same conclusion using daily oil price data. Studies of historical volatility behaviour based on weekly crude oil data, along with weekly oil products data and futures prices since the early 1990s to 2005, confirm the case for volatility, especially in the post-2000 period (Kang and Yoon 2009). Other studies, such as that of Labban (2010), also suggest an increase in the volatility of the oil spot prices.

The issue of increasing volatility levels dominates all discussions of the financialisation of the oil market. The debate is not settled, however, as different data and different calculation methods – perhaps not surprisingly – yield different results. In opposition to the studies surveyed above, for example, this research is based on day-to-day volatility data, which fail to indicate any volatility increase at all. On different grounds, Daniel Yergin, author of *The Prize: The Epic Quest of Oil, Money and Power* (2008), argues that volatility levels in the oil market have only decreased in the wake of the growth of the paper market. Additionally, it suggested that the hedge funds operating in commodity markets are not perpetrators of destabilising volatility, as much as purveyors of liquidity (Haigh et al. 2005).

This book maintains that the volatility levels of the oil market have been gradually increasing. Hence, the measurement of day-to-day volatility rates cannot be considered as a reliable methodological tool for studying the functioning of commodity markets where actor behaviour is shaped around protracted international political and economic developments. These developments may be anything from social or military turbulence to revised international growth rates, production cuts or new scientific discoveries. Moreover, the argument of this book shows that volatility rates individually are not as relevant to the financialisation of the oil market as the combination of volatility rates and inflated price levels. This is because the same rate of volatility has very different economic and financial effects in the internal and external aspect of a market with a $30 price level at one end and $120 at the other, especially in the case of the oil market, where production and refinery costs remain inelastic to changes in the price level of the commodity.

Tables 10.3 and 10.4, produced by the EIA, trace the correlations between the price of WTI crude oil and the major commodities

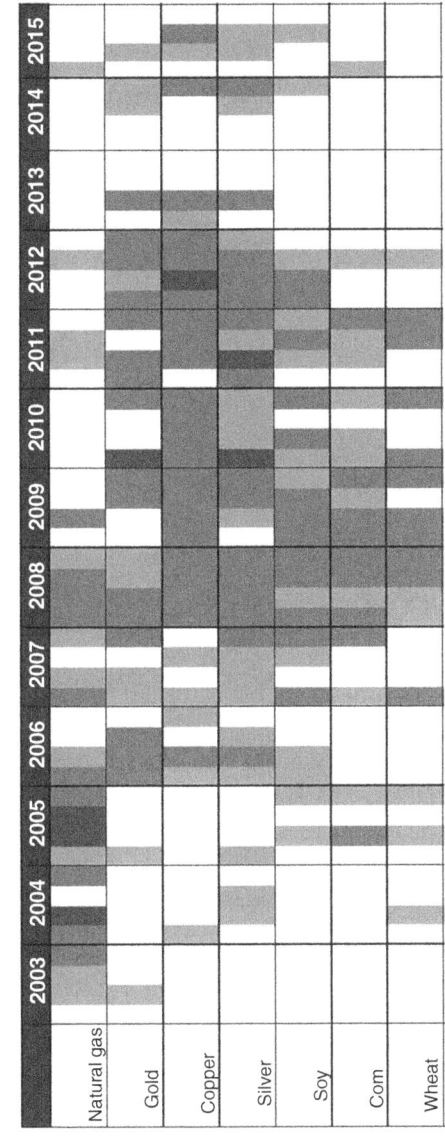

Table 10.3 Correlations between daily price changes of crude oil and other commodities (EIA Data 2016)

Table 10.4 Correlations between daily price changes of crude oil and oil futures and financial investments (EIA Data 2016)

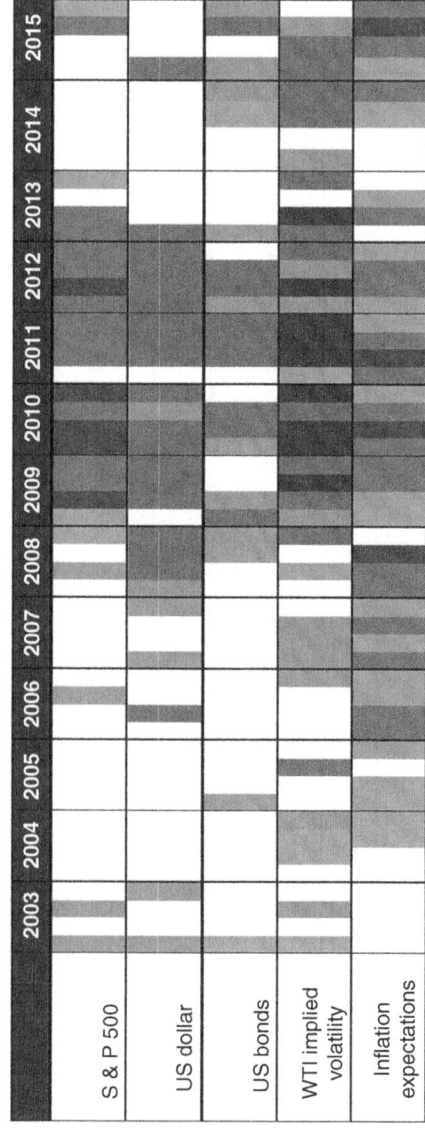

Note: Correlations computed quarterly

throughout the last decade, in the form of natural gas, gold, silver, soybeans, corn and wheat. It also traces the correlations between crude oil futures returns and financial investments in US dollars, the S&P 500 and US government bonds. This table illustrates, on the one hand, a growing correlation among the price levels of these commodities as a result of their indexification and, on the other hand, their negative correlation to the performance of the stock markets after 2008, which spelled the end for their use as hedging investments and, with it, the high liquidity levels they attracted during the early 2000s. This can also be regarded as a result of the financialisation of the market, as the links between the markets grow, while the spillover effects from one market to the other intensify.

References

Behr, T. (2009). The 2008 oil price shock competing explanations and policy implications. *GPPi Global Energy Governance Project Policy Paper Series*.

Bekiros, S., & Diks, C. (2008). The relationship between crude oil and futures prices: Cointegration, linear and nonlinear causality. *Energy Economics*, 30, 2673–2685.

Bloomberg Data. (2015). *Data terminal*. London: Bloomberg. Accessed 14 October 2015.

Büyükşahin, B., Haigh, M., Harris, J., Overdahl, J., & Robe, M. (2008). Fundamentals, trader activity and derivative pricing. *US Commodity Futures Trading Commission*.

Campbell, J., & Viceira, L. (2003). A multivariate model of strategic asset allocation. *Journal of Financial Economics*, 67(1), 41–80.

CFTC. (2008). *Staff report on commodity swap dealers & index traders with commission recommendations*. Commodity Futures Trading Commission.

CFTC. (2009). *About the commitments of traders reports*. Commodities Futures Trading Commission.

Chong, J., & Miffre, J. (2010). Conditional correlation and volatility in commodity futures and traditional asset markets. *Journal of Alternative Investments*, 12(3), 61–75.

Choy, M., Gregorio, D., & Burgdorfer, B. (2012, September 19). Oil dives $4 as supplies rise, Saudi talk spooks funds, *Reuters*.

Cifarelli, G., & Paladino, G. (2010). Oil price dynamics and speculation: A multivariate financial approach. *Energy Economics*, 32(2), 363–372.

Duffie, D., Gray, S., & Hoang, P. (2004). Volatility in energy prices. In V. Kaminski (Ed.), *Managing energy price risk*. London: Risk Publications.

EIA Data. (2016). *EIA Data*. Energy Information Administration.

Fattouh, B. (2011). An anatomy of the crude oil pricing. *Oxford Institute of Energy Studies.*
Frankel, J., & Rose, A. (2009). Determinants of agricultural and mineral commodity prices. *Kennedy School of Government, Harvard University.*
Gros, D. (2008), July 9. The China bubble fuelling record oil prices. *Financial Times.*
Haigh, M., Hranaiova, J., & Overdahl, J. (2005). Price dynamics, price discovery and large futures trader interactions in the energy complex. *US Commodity Futures Trading Commission.*
Irwin, S., Sanders, D., & Merrin, R. (2009). Devil or angel: The role of speculation in the recent commodity price boom (and bust). *Journal of Agricultural and Applied Economics*, 41(2), 377–391.
Kang, S., & Yoon, S. (2009). Forecasting volatility of crude oil markets. *Energy Economics*, 31(1), 119–125.
Kaufmann, R. (2011). The role of market fundamentals and speculation in recent price changes for crude oil. *Energy Policy*, 39(1), 106–115.
Kaufmann, R., & Ullman, B. (2009). Oil prices, speculation, and fundamentals: Interpreting causal relations among spot and futures prices. *Energy Economics*, 31(4), 550–558.
Khan, M. (2009). The 2008 oil price 'Bubble'. *Peter G. Peterson Institute for International Economics.*
Kyle, A., & Xiong, W. (2001). Contagion as a wealth effect. *Journal of Finance*, 56(4), 1401–1440.
Labban, M. (2010). Oil in parallax: Scarcity, markets, and the financialization of accumulation. *Geoforum*, 41(4), 541–552.
Markowitz, H. (1952). Portfolio selection. *Journal of Finance*, 7(1), 77–91.
Masters, M., & White, A. (2008). The accidental hunt brothers: How institutional investors are driving up food and energy prices. *Masters Capital Management and White Knight Research and Trading.*
Masters, M., & White, A. (2009). The 2008 commodities bubble: Assessing the damage to the United States and its citizens. *Masters Capital Management and White Knight Research and Trading.*
Meyer, L. (2001). Inflation targets and inflation targeting. *BIS Review*, 65, 1–13.
Parsons, J. (2010). Black Gold & Fool's Gold: Speculation in the oil futures market. *Economia*, 10(2), 81–116.
Pindyck, R., & Rotemberg, J. (1990). The excess co-movement of commodity prices. *The Economic Journal*, 100(403), 1173–1189.
Regnier, E. (2007). Oil and energy price volatility. *Energy Economics*, 29(3), 405–427.
Silvapulle, P., & Moosa, I. (1999). The relationship between spot and futures prices: Evidence from the crude oil market. *Journal of Futures Markets*, 19, 157–193.

Silvennoinen, A., & Terosvirta, T. (2009). Modelling multivariate autoregressive conditional heteroskedasticity with the double smooth transition conditional correlation GARCH model. *Journal of Financial Econometrics*, 7(4), 373–411.

Silvério, R. (2010). The role of financial agents in 2006–2008 oil price rise: Evidence from new commitment of traders methodology. In *33rd International Association of Energy Economics Congress*. Rio de Janeiro.

Tang, K., & Xiong, W. (2009). Index investing and the financialization of commodities. *Princeton University*.

Tavasci, D., & Ventimiglia, L. (2011). Financialisation in the Primary Commodity Dependent Developing Countries: The Case of Chile. *International Journal of Management Concepts and Philosophy*, 5(2), 171–189.

Tokic, D. (2011). Rational destabilizing speculation, positive feedback trading, and the oil bubble of 2008. *Energy Policy*, 45, 6009–6015.

UK Cabinet Office. (2008). *The rise and fall in oil prices: Analysis of fundamental and financial drivers*. London: UK Cabinet Office.

UNCTAD. (2009). *Trade and development report 2009*. New York: United Nations Publications.

United States Senate. (2006). *The role of market speculation in rising oil and gas prices: A need to put the cop back on the beat, committee on Homeland Security and governmental affairs*. Washington, DC: Staff Report of US Senate Permanent Subcommittee on Investigations.

Wigan, D. (2008). A global political economy of derivatives – Risk, property and the artifice of indifference. *Brighton University*.

Wray, L. (2009). The rise and fall of money manager capitalism: A Minskian approach. *Cambridge Journal of Economics*, 33, 807–828.

Yergin, D. (2008). *The Prize: The Epic Quest for Oil, Money & Power*. New York: Free Press.

CHAPTER 11

Epilogue

Abstract This chapter argues that in an attempt to offer a fresh approach to the political economy of the oil market, this book traced the evolution of the oil market through the three key phases of the financialisation process and identified the effects of the latter on the structure and behaviour of both the oil market and the US macroeconomic and financial performance during the three most recent oil shocks. Even though the purpose of this book was not to provide a new theoretical approach on either crises or financialisation, it demonstrated and traced the manifestation of this process in the oil market, which has long been overlooked in the literature on financialisation. In doing so, it brought to light the increasingly important role that the oil market has had on the contemporary political and economic history, particularly its role in the formation and development of the 2008 credit crisis, as well as its potential for shaping further political–economic turbulence in the future.

Keywords Oil market · Oil shocks · Economic history · Financial performance

The analysis conducted in this book set out to answer a number of questions and ended its journey with as many conclusions and findings. Dismissive of the peak-oil arguments typically cited in the context of oil market evolution, this book suggested that the evolution of oil market was driven and conditioned by the process of financialisation that started some three decades ago. Even though the academic efforts to conceptualise the process of financialisation are quite recent, particularly with reference to the commodity markets, the developments that have taken place in the oil market, and that have altered its structure, motives, fundamental function and processes, are unambiguously symptomatic of the financialisation phenomenon. The direct causal link between these developments and the evolution of the financialisation of oil markets was further evidenced by their concomitance with the deregulation of the oil market and market's opening to the financial sector.

The starting point for the analysis was therefore setting the basic framework for evaluating and interpreting (a) the development and effects of the financialisation process in the oil market structure and (b) the effects of the relationship between the financialisation process and the macroeconomic and financial performances. The notion of financialisation proposed by Epstein (2005, p. 3) formed the basis for the approach employed in this study, maintaining that the process of financialisation "means the increasing role of financial motives, financial markets, financial actors and financial institutions in the operation of the domestic and international economies". At the same time, the analysis in this book gave particular attention to the regulatory and technological developments that impacted the oil market's evolution. Most importantly, the study placed its focus on the change in the actors' structure within the oil market and the role of financial actors in influencing the market dynamics and fundamentally changing its modus operandi.

To that end, one may highlight three main conclusions and findings of this book. First, it demonstrated how a new group of actors appeared in the triangular structure of the oil market during the 1980s with the introduction of oil-based financial products. This introduction opened up the market and allowed for the participation of financial investors, who eventually became an integral part of the oil market structure. It didn't take long for the financial actors to grow exponentially in numbers and in influence, thanks to the two key events. The first was the development of commodity-based indexes in the early 1990s (Clapp and Helleiren 2010); the second was the introduction of the CFMA regulation and the consequent precipitous inflow of speculative capital in the market in the early 2000s (Brown-Hruska 2004). At that point, a cleavage within the new group of market actors emerged, dividing traditional financial investors with hedging interests and their speculative, or price-manipulative, counterparts.

The second finding maintains that, over time, this new group of actors has grown, to include more and more players, in stark contrast to the underlying oil commodity market actors whose number has been relatively stable over the period under study. Evidence of this is the fact that the value of trades in the financial dimension of the oil market outstripped that of the physical commodity market (Masters and White 2009), even in estimates that exclude the value of OTC trades, which cannot be measured reliably (UNCTAD 2009). This growth took place against the background of the deregulation of the market, the development of new technologies, the introduction of new financial products, and the use of the Internet in the marketisation of these products, which made the market easily accessible to a growing pool of international investors.

Finally, the analysis in this book demonstrated that the increase in volatility in the WTI spot price during the market's financialisation period under study could be regarded as a direct effect of the investment behavioural patterns of the new financial actors (Clapp 2009). As the oil futures market performance was linked to that of the spot prices by virtue of the market structure, the emergence of a fourth group of financial actors (interested in either rising or changing oil prices) naturally led to increasing volatility of oil spot prices. In addition to that trend, an important correlation between the newly introduced volatility in the oil market performance and the performance of the international asset markets was unveiled (Tang and Xiong 2009).

The evolution of the oil market structure (which arguably took place through the financialisation process) was traced along with its reciprocal relationship with the international economy and the US economy, in particular. To that end, employing the concept of oil shocks was instrumental for revealing the effects of financialisation process in the oil markets on macroeconomic dynamics, with pre-financialisation shocks taken as a baseline. This analysis, in turn, helped identifying the indicators and the effects of the symbiotic relationship between the financialisation process in the oil markets and macroeconomic performance. A close reading of the literature and the data on the evolution of the oil market in the context of the international economy, and that of the USA specifically, indeed suggests strong links – and the evolving relationship – between the oil market and macroeconomic performance (Mork 1989). When examining such links, it became evident that the most pronounced are those between the oil market and productivity levels through input costs, real wage levels, inflation levels, unemployment levels, net investment levels, exchange rates and, very importantly, budget balances. What is more, the interplay between the two market dynamics proved to have some spillover effects on the relationship between the oil market and the asset market.

Having established the relevant analytical framework and having refined the approach of this study, it moved onto a historical review of the three phases of the financialisation of oil markets, employing three major oil shocks of the post-1980s era respectively, and using them as barometers of the financialisation process and its effects. These phases were therefore framed into a chronological triptych of low (1989–1999), early (1999–2003) and advanced (2002–2008) financialisation. Even though the three oil shocks under study, namely the ones in 1991, 2001 and 2008, differ in a number of respects (Blanchard and Galí 2008), their close proximity to the macroeconomic crises and financial downturns that took place in those periods, proved to be particularly useful and illustrative for unfolding the financialisation phenomenon. For they helped setting the relevant macroeconomic context of the financialisation process in oil markets and reveal the drivers of the events around the oil shocks and the subsequent economic crises. This way, it became possible to interpret the links and causalities between the development of the oil market throughout these three phases and the relevant macroeconomic dynamics. In what follows, the key characteristics of the respective phases will be briefly outlined.

The analysis of the 1991 oil shock and economic and financial turbulence shows that this period coincides with the origins of the emergence of the financialisation process in the oil market. This is because short, non-commercial, and otherwise risky investments started becoming increasingly popular among purely financially driven investors who were mostly positioned outside of the oil market structure until 1992 (Silvennoinen and Terosvirta 2009). This analysis suggests that even though oil-based financial products had already been available for a decade, financial actors did not start entering the market before 1992. In 1992, the introduction of commodity-based indexes made the market appealing and approachable to purely financial, non-oil-related, investors who could not have previously entered the market as easily (Tang and Xiong 2009). This, however, meant that the 1991 oil shock and the subsequent macroeconomic and financial crises did not have any financialisation-derived links as they both occurred prior to 1992. To be sure, this does not preclude the existence of traditional links between these three entities, especially since the oil market performance had an influential role in the US macroeconomic and financial performance of this period.

Even though the 1991 oil crisis can be regarded as a purely market-driven crisis, its effects were shortly passed on to the weak and unstable US economy. The analysis based this argument on a certain level of synchronicity and causality that can be observed between the behaviours of these three entities during this period. It is then suggested that this causal relationship is based on the US dollar marketisation of oil, which, combined with the pre-existing large budget deficits, weakened the external aspect of the US economy (Clapp and Helleiren 2010). At the same time, the ensuing deterioration of the main macroeconomic indicators, and the fear of a third large oil shock, affected consumer confidence levels (Roubini and Setser 2004). That fear, coupled with unfavourable interest and inflation rates, led to a crash in the financial markets. The absence of high volumes of investments, especially speculative investments, indicates that these links were not influenced by the effects of the financialisation process of the oil market which only started months later; however, they do stress the influence that the oil market performance had on the US macroeconomic and financial performance even before the emergence of the financialisation process. Therefore, this period is defined in terms of low financialisation, as the roots of the process can be traced within its boundaries, even though the crisis itself was not shaped by it. As a result, the study of this crisis provides us with a benchmark of the effects of an oil shock in the absence of this process.

The early financialisation period, in turn, was analysed through the 2001 oil shock and its macroeconomic consequences. Similarly to the 1991 shock, this shock was also preceded by a long booming economic period, and an attempt by the Federal Reserve to slow down the economy through successive interest rate increases in fear of corrosive inflation (Cooper 1992). At the same time, the election of Chávez, and his OPEC leadership aspirations, had inflated the oil price level (Rodríguez 2002), which would only increase further with the invasion of Afghanistan on October 2001. Thus, this macroeconomic policy, combined with a reduction in the US taxation levels, financed through the budget surplus and the oil market performance, led to an economic recession in the beginning of 2001. The effects of the macroeconomic performance were soon passed on to the financial performance, thus bursting the dot-com bubble, and crashing the 'new economy' ideas along with it. The combination of these two crashes and the oil shock led to a wide erosion of the US macroeconomic performance, causing high unemployment, negative growth and increased inflation as well as a significant loss of confidence across the USA.

The period surrounding the 2001 oil shock was indeed of great importance for the future development of the financialisation of the oil market. More specifically, three developments altered the environment of the oil market, facilitating the process of financialisation in its structure: (1) the end of the 'prudent investor' rule, (2) the development of Internet-based trading and (3) the CFMA regulation. These developments allowed large financial institutions to enter the oil-based financial market through the back door of the OTC market (Brown-Hruska 2004). At the same time, the introduction of Internet-based trading, combined with the development of the commodity-based indexes that had been launched in the previous period, allowed for a massive inflow of capital in the market from any international financial investor and with little to no financial or physical barriers (Labban 2010).

In other words, these three developments allowed for the full participation of a fourth actor in the triangular oil market structure. It becomes evident from the performance of the oil-based financial products prior to the peak of this shock in 2001, which is synchronised with the performance of the oil spot price. In contrast, a decoupling of this performance can be observed in the post-2000 period, which allows for the futures products to retain most of their value as the oil spot price plundered.

Even though the effects of financialisation can be best observed in the post-2000 period, its influence in the built-up to the 2001 crisis is nevertheless apparent. The macroeconomic downturn of the USA was a direct result of the dot-com crisis, which, along with the rising oil price level and the terrorist attacks, led to a crash of the US stock markets and the loss of confidence in the USA and the 'new economy' paradigm (Arestis and Karakitsos 2004). The military activities that took place in the Middle East, shortly after the above events, was an indicator that any investment in the oil market would become profitable, especially since the asset markets were underperforming and the oil-based financial products had proved to serve well as an alternative to the asset markets and the inflation pressures (UNCTAD 2009). In light of the 2000 regulatory developments, investments in the oil market became both appealing and accessible to a wider audience of investors. Consequently, while the asset markets were plunging and the macroeconomic performance was still in a recessionary period, the oil-based financial products dropped marginally and for a very short period of time after the peak of the 2001 shock, only to return promptly back to their inflationary trend.

In contrast to the previous phases, during advanced financialisation the effects of the financialisation process in the oil market and its links to the US macroeconomic and financial performance had become a very real – and integral – part of the system. Very similar to the previous two shocks, the US economy was very unstable in the months preceding the 2008 shock, which marks this financialisation period. However, this instability was of a different nature. The Federal Reserve, in its attempt to boost the economy after the events of 2000, had maintained very low interest rates and, in doing so, had boosted the level of liquidity of the economy and shaped the subprime mortgage bubble (Hardouvelis and Stamatiou 2009). At the same time, two very important underlying developments were shaping the events of the period. The first of these developments was the increase in US deficits. This increase had been caused by the loss of value of the US dollar, the unforeseen fiscal expenditure of the military activities in the Middle East and the ever-increasing price of oil. The second development was the exponential increase in the inflow of financial capital in oil-based financial markets.

Following to that, the analysis of the 2008 shock identified a causality link between the relevant economic phenomena in this period. The low interest rates introduced in 2001 boosted the housing market, leading to the emergence of a bubble; simultaneously, oil market was experiencing an increase in financial activity as well, with the only difference being that, in the case of the latter, this increase was more gradual and was intensified in 2007, by the time the housing market started deflating. It is suggested that the turn of the financial capital from the housing market to the oil market, as a result of the deflation of the bubble in the housing market, was synchronised with the increase in activity in the oil market. The relevance of this claim is supported by the fact that, when the housing bubble burst and the real estate prices kept falling for the following years, the oil market maintained its upward rally for more than a year before dropping back to its 2003 levels. This dynamic was permitted by the self-fulfilling prophecy that was set in motion by the market expectations of a steady rise in oil price levels (due to the military activities in the Middle East), supply shortages and reduced spare-capacity levels.

Since 2007, when the price level of oil began to rise, the effects of this hike had a direct impact on the macroeconomic performance of the USA.

The increased demand for oil-based products created additional inflationary pressures on the spot price level of oil, causing an exacerbation of the US budget deficits, because they now had to purchase oil for a higher price and with a weaker currency. This shaped pressures in both the external and the internal aspect of the US macroeconomic performance. The levels of output and inflation deteriorated as the cost of energy rose, which in turn increased unemployment and placed pressures on the economy starting from end of 2007.

These pressures were also intensified by the effects of the crashing housing market, which was not only reducing the liquidity of the economy (Portes et al. 2009), but also led to a profound loss of confidence among the US public. The performance of the asset market followed the performance of the macroeconomy very closely. As the housing market bubble started failing, a number of large financial institutions faced liquidity problems, which in some cases led to near, or actual, bankruptcies. The effects of these events, combined with the deteriorating macroeconomic performance and the already negative confidence levels, led to a crash of the asset markets and a flight of capital. In 2008, an official macroeconomic recession was announced just as the oil price was reaching its historical peak.

The analysis of the events of this crisis, therefore, brings to light the effects of the process of financialisation when taking into consideration the importance of the nature and accessibility of the oil market in the development of this crisis. On the one hand, the nature of oil as a primary commodity with inelastic international demand meant that the changes in its price levels were fully experienced by the USA, as an importing economy; insofar as they could not adjust their demand level is a timely manner. On the other hand, the introduction of uncomplicated commodity-based index products in the 1990s, the deregulation of the oil-based financial products in 2000 and their easy accessibility through the development of new technologies shaped a market that attracted an oversized and progressively increasing amount of financial capital after 2001. However, due to its nature, the growth of this market created negative pressures on the performance of both the US economy and its financial markets.

The first part of the final Part III of the book placed its focus on the protagonists and drivers of the financialisation process in the oil markets – its actors and the new financial actors in particular. The synthesis of the conclusions reached throughout the study allowed for conceptualising and describing the new four-actor oil market structure, which emerged with the establishment of the financial segment of the market, thanks to the relevant financialisation phenomena. This, in turn, laid the grounds for the second part of the final chapter, which discussed the consequences of the establishment and integration of the new financial actors in the oil markets, having in mind their interests, strategy, tactics and behaviour. The interplay between the actors in the physical segment of the oil market and the financial one subsequently revealed the reasons behind the process of decoupling of asset markets and commodity markets within the oil market structure (as well as its effects). This decoupling was further evidenced through the study of the relationship between oil spot prices and oil futures, which explains the growing price volatility and the behaviour of financial market players. In other words, the final part provided insights into the way the financialisation process shaped the structure of the present oil market and defined the role of the financial actors within it.

This book indeed aimed at guiding the reader through the maze of economic theories and approaches to conceptualising the oil market's evolution and its link with macroeconomic dynamics, ultimately offering a map with topography of the contemporary oil market structure, as well as a framework for understanding its relationship with international macroeconomic dynamics. In pursue of this, the present study explained the transformation that the oil market has undergone since the 1980s and its reciprocal link with macroeconomic performance of the international markets and that of the USA specifically; while at the same time, offered insights into the functioning of the oil market, as we know it today.

This way, in an attempt to offer a fresh approach to the political economy of the oil market, this book traced the evolution of the oil market through the three key phases of the financialisation process and identified the effects of the latter on the structure and behaviour of both the oil market and the US macroeconomic and financial performance during the three most recent oil shocks. The purpose of this book was not to provide a new theoretical approach on either crises or financialisation. It rather intended to demonstrate and trace the manifestation of this process in the oil market, which has long been overlooked in the literature of financialisation.

In doing so, it brought to light the increasingly important role that the oil market has had on the contemporary political and economic history, particularly its role in the formation and development of the 2008 credit crisis, as well as its potential for shaping further political–economic turbulence in the future. Given the nature of oil as a naturally exhaustible resource, and a primary energy source, the process of financialisation and decoupling from its physical production makes the potential for further far-reaching oil shocks very high. In this sense, understanding the functioning of the oil market in the context of the evolutionary pattern that it is conditioned by, as well as its relationship with the international macroeconomic and financial performance, is crucial. For it allows to look beyond the short-term market volatility, and to identify the relevant market dynamics and long-term trends.

REFERENCES

Arestis, P., & Karakitsos, E. (2004). *The post bubble US economy: Implications for financial markets and the economy.* London: Palgrave Macmillan.

Blanchard, O., & Galí, J. 2008. The macroeconomic effects of oil price shocks: Why are the 2000s so different from the 1970s? In J. Galí & M. Gertler *International dimensions of monetary policy.* Chicago: University of Chicago Press.

Brown-Hruska, S. (2004). Securities Industry Association Hedge Funds Conference. CFTC.

Clapp, J. (2009). Food price volatility and the vulnerability in the global South: Considering the global economic context. *Third World Quarterly, Routledge.*

Clapp, J., & Helleiren, E. (2010). Troubled futures? The global food crisis and the politics of agricultural derivatives regulation. *Review of International Political Economy,* 19(2), 181–207.

Cooper, R. 1992. The Middle East and the world economy. In J. Nye & R. Smith *After the storm, lessons from the Gulf War.* London: Madison Books.

Epstein, G. (2005). *Financialization and the world economy.* London: Edward Elgar.

Hardouvelis, G., & Stamatiou, T. (2009). Hedge funds and the US real estate bubble: Evidence from NYSE real estate firms. *University of Piraeus.*

Labban, M. (2010). Oil in parallax: Scarcity, markets, and the financialization of accumulation. *Geoforum,* 41(4), 541–552.

Masters, M., & White, A. (2009). The 2008 commodities bubble: Assessing the damage to the United States and its citizens. *Masters Capital Management and White Knight Research and Trading.*

Mork, K. (1989). Oil and the macroeconomy when prices go up and down: An extension of Hamilton's results. *Journal of Political Economy,* 97(3), 740–744.

Portes, R., Dewatripont, M., & Freixas, X. (2009). Macroeconomic stability and financial regulation. *CEPR,* 178.

Rodríguez, A. (2002). *Saramago: Soy Un Comunista Hormonal.* Venezuela: Capital Intelectual.

Roubini, N., & Setser, B. (2004). The effects of the recent oil price shock on the U.S. and global economy, Stern School of Business NYU and global economic governance programme. University College Oxford.

Silvennoinen, A., & Terosvirta, T. (2009). Modelling multivariate autoregressive conditional heteroskedasticity with the double smooth transition conditional correlation GARCH model. *Journal of Financial Econometrics*, 7(4), 373–411.

Tang, K., & Xiong, W. (2009). Index investing and the financialization of commodities. Princeton University.

UNCTAD. (2009). *Trade and development report 2009*. New York: United Nations Publications.

INDEX

A
Accumulation, 28
Accumulation regime, 32
Actor behaviour, 44, 50, 52, 53, 54, 57, 132
Actor structure, 109
Adolf Hitler, 19
Affordability, 14
Afghanistan, 97, 143
Alternative energy, 14, 19, 131
Andrew Mellon, 14
Anglo-Persian Oil, 15
Arab-Israeli War, 16, 18
Asset markets, 68, 69, 112, 122, 123, 131, 141, 144
Automotive, 95
Autonomisation, 28
Autumn, 29
Availability, 14, 33, 50

B
Bailout, 96
Bank of England, 95
Banks, 18, 25, 33, 48, 71, 90, 94, 95, 101, 113, 118, 123
Bargaining, 16, 127
Bear Stearns, 95
Behavioural finance, 52

Ben Bernanke, 95
Bill Clinton, 25
BNP Paribas, 118
BP, 15
Breakthroughs, 14, 56
Brent Crude, 80
Bretton Woods, 24
Bubble, 51, 52, 58, 61, 64, 65, 66, 68, 87, 88, 93, 98, 102, 103, 104, 110
Budget deficit, 18, 71, 78, 80, 89, 95
Budget deficits, 72, 89
Business cycles, 42
Buy, 67, 100, 127

C
Capital, 24
Capital accumulation, 32
Capitalist accumulation, 27
Capital losses, 111
Cartel, 15, 21
CCP, 118
Central Counter Party, 118
CFMA, 89, 90
CFTC, 67, 74, 82, 89, 100, 103, 112, 118, 121, 131
Chicago school, 50
Cognitive evaluations, 46

INDEX

Commercial, 25, 82, 83, 98, 100, 102, 113, 118, 120, 129, 130
Commercial investments, 82
Commodity derivatives, 40, 124
Commodity markets, 40, 68, 69, 81, 99, 121, 122, 126, 131, 132, 140, 147
Communications, 57
Co-movement, 122
Concession agreements, 15
Conspicuous consumption, 51
Consumer confidence, 73, 80, 88, 96
Consumers, 59, 68, 81, 103, 118, 124
Consumer spending, 32
Contagion, 48
Copper, 130
Corn, 135
Corrosive inflation, 87
Crude oil, 15, 17, 19, 20, 67, 89, 98, 100, 113, 118, 126, 127, 128, 129, 131, 135

D

Decoupling, 24, 27, 30, 31, 42, 48, 69, 94, 123, 130, 144
Deregulation, 25, 58, 112, 131
Destabilising volatility, 132
Deutsche Bank, 81
Devaluation, 18
Dispersion hypothesis, 71
Diversification, 29, 66, 68, 81, 122
Dodd-Frank Act, 113
Dollar, 18, 19, 25, 80, 97, 100, 122, 143
Donald H. Rumsfeld, 97
Dot-com, 87, 88
Dow Jones, 81

E

Economic Behaviour, 52
Economic geography, 29
Economic imperialism, 26
Efficiency Market Theory, 49
EIA, 89, 132
Electronic marketisation, 56
Elites, 31
Embargo, 18, 78
Employment, 72, 94
EMT, 49, 50
Energy market, 2, 121
Equity shocks, 122
Exchange rate, 72, 80, 82, 100, 101, 104
Expectations, 40, 41, 44, 45, 46, 48, 49, 51, 52, 53, 54, 55, 72, 73, 79, 80, 83, 96, 99, 101, 102, 103
Extraction, 16, 66
Exxon, 15

F

Fannie Mae, 95
Federal Budget, 77
Federal Reserve, 18, 25, 43, 78, 87, 95, 102
Feedback loop, 128
Feelings, 46, 53, 54, 96
Fiasal, 19
Finance-led capitalism, 26
Financial actors, 24, 40, 41, 48, 49, 52, 53, 54, 55, 57, 58, 59, 66, 68, 89, 90, 99, 111, 112, 123, 125, 128, 129, 131
Financial asset, 68, 81, 97, 110, 122
Financial crisis, 80, 94, 96, 99, 102, 103
Financial innovation, 67, 87, 90
Financialisation, 20, 22, 24, 25, 26, 27, 28, 29, 30, 35, 39, 49, 54, 55, 58, 60, 65, 66, 67, 70, 71, 73,

83, 89, 98, 104, 109, 110, 111, 112, 122, 123, 124, 130, 131, 132, 135
Financialised accumulation, 67
Financial liberalisation, 33
Financial market, 25, 33, 40, 44, 47, 49, 50, 51, 52, 68, 73, 94
Financial motives, 24, 34, 83
Financial products, 27, 40, 56, 58, 67, 82, 89, 91, 93, 95, 101, 121, 125, 128
Financial profitability, 33
Financial profits, 27, 30, 59
Fordism, 33
Foreign exchange, 66
France, 16
Freddie Mac, 95
Fundamental economic indicators, 73, 79
Fundamentals, 52, 55, 69, 73, 86, 110, 126, 128, 129, 130
Futures, 7, 21, 66, 67, 68, 73, 80, 82, 83, 89, 90, 97, 112
Futures speculation, 121

G
GDP, 20, 43, 72, 86, 103
Gender Trouble, 44
George Bush, 77
Glass-Steagall Act, 25
Globalisation, 29, 39, 71, 96
GNP, 94, 96
Gold, 18, 24, 81, 101, 113
Goldman Sachs, 81, 113
Gold standard, 24
Gordon Brown, 121
Government expenditure, 32, 78
Government intervention, 34
Great Depression, 18, 24
Gross Domestic Product, 43

GSCI, 81, 122
Gulf, 14, 15, 79

H
Hedge funds, 90, 91, 96, 101, 113, 118, 132
Hegemonic, 19, 29
Herding, 53, 54, 57, 122
Heuristics, 46, 53, 54, 73
Homo economicus, 45
Homogenisation, 33
Hostages, 19
Hostile takeovers, 47
Housing market, 93, 102, 104
Hugo Chávez, 88
Hydro, 14

I
ICE of London, 120
Importing economies, 72, 146
Imports, 17, 19, 58, 72, 104
Index, 43, 68, 80, 81, 86, 96, 98, 101, 103, 112, 113, 120, 122, 124, 131, 146
Indexification, 135
Inelastic, 14, 72, 78, 101, 110, 121, 132, 146
Inflation rates, 44, 71, 79
Information revolution, 57
Institutional, 26, 27, 29, 31, 32, 33, 34, 40, 41, 83, 94, 129
Institutional approaches, 31
Institutionalisation, 32
Interest rates, 43, 71, 73, 79, 87, 95, 102, 145
International Oil Companies, 112
International Petroleum Exchange, 21, 66
Internet, 56, 57, 67, 89, 111, 123, 141

Inventory crisis, 88
Investment, 14, 25, 27, 29, 33, 34, 35, 40, 43, 44, 47, 48, 50, 51, 53, 54, 65, 66, 67, 68, 71, 72, 81, 82, 83, 88, 90, 94, 95, 97, 98, 100, 101, 104, 110, 113, 120, 121, 129, 130
Investment theory, 43
IOCs, 112
IPO, 86
Iran, 15, 17, 19
Iraq, 15, 18, 78, 96, 97, 101, 104

J
Japan, 17
JP Morgan, 113, 118

K
Keynes, 50, 51
Kissinger, 17
Kuwait, 15, 78

L
Laissez-faire, 50
Lehman Brothers, 96
Lenin, 26
Liquidation, 48
Liquidity, 34, 46, 48, 57, 58, 94, 95, 126, 135, 145, 146

M
Macroeconomic equilibrium, 47
Macroeconomic indicators 42, 44, 129
Macroeconomic performance, 44, 46, 48, 70–72, 79, 80, 89, 96, 102, 104, 141, 143–147

Manufacturing, 86, 88
Margaret Thatcher, 25
Marketing, 14–16, 66, 123
Marketisation, 16, 71, 112, 131
Market structure, 83, 109, 111, 123
Marxian, 28
Marxist, 26, 34
Mediators, 122, 124
Mediterranean, 16
Mental calculations, 53
Merrill Lynch, 118
Middle East, 15–18, 20, 78, 95, 102, 104, 128
Military operations, 79
Military products, 17
Military spending, 97
Minsky, 47, 51, 52, 54
Mob psychology, 52–55, 57
Modernization Act of 2000, 89
Monetary policy, 47, 71, 78, 79, 91, 95
Money fetish, 29
Money growth, 28
Moneylenders, 34
Monopoly, 21, 27
Morgan Stanley, 118
Motives, 34, 41, 65–66, 98, 120, 123
Multinationals, 16, 28

N
National Oil Companies, 112, 113
Natural gas, 14, 67, 135
Neo-classical, 50
Netherlands, 16, 18
New economy, 86, 88
New products, 66
Nicolas Sarkozy, 121
Nigeria, 16
Nixon, 17, 18
NOCs, 113, 118, 131

Noise-trade strategies, 131
Non-commercial 82, 99, 100, 102, 113, 118, 120, 121, 130
Non-commercial investments, 82
Non-communist, 15
NYMEX, 73, 74, 98, 99, 119, 120, 125

O

Occidental, 15
Oil crisis, 66, 110
Oil futures, 66, 73, 74, 80, 82, 89, 90, 97–111, 113, 118, 120, 121, 123, 125, 128, 130, 135
Oil production, 15, 17, 18, 19, 78, 81
Oil reserves, 19, 78, 79, 128
Oil shocks, 70–73
Online platform, 126, 127
OPEC, 15–17, 19–21, 78, 88, 99, 102, 110, 128
Options, 67
Organisation of Petroleum Exporting Countries, 15
OTC, 67, 88–90, 98, 99, 118, 124, 131
Over-accumulation, 29, 30
Over the counter, 67
Overproduction, 28

P

Paper market, 70, 99, 132
Paradox of thrift, 51
Parasitical rentiers, 26
Pension funds, 48, 68, 90, 91, 113
Performative cycle, 40, 48, 55, 127, 128
Performative economic theory, 123
Performativity, 42, 44, 45, 48
Phillips Curve, 86
Platts, 126–128

Positive reinforcement, 87
Posted price, 15, 16, 19
Post-Fordist, 24
Post-Keynesian, 34, 35, 49–52, 54
Price determination, 40, 49, 127
Price manipulation, 110, 131
Price manipulators, 34, 131
Price-setting mechanism, 49, 50, 112, 125, 126, 129
Price shocks, 71
Producers, 19, 59, 66, 68, 81, 111, 112, 118, 120, 124, 126, 128
Productivity, 18, 29, 32, 33, 46, 48, 72, 95, 141
Profitability, 16, 26, 30, 57, 86, 97
Profit creation, 31
Prudent investor, 68, 90

R

Real economy, 30–32, 34, 42, 45, 48, 49
Real estate, 80, 95
Real production, 24, 27, 28, 30, 31, 34, 35, 59, 121
Recession, 18, 28, 78, 88, 94, 96, 104
Reconfiguration, 21
Redistribution, 20, 28, 35, 72
Refining, 15, 17, 66, 142
Reflexivity, 53
Regulation, 28, 32–35, 55, 56, 58, 59, 89–91, 103, 113, 124
Regulation School, 31–33
Regulatory developments, 55, 58, 59, 66
Rentiers, 34
Reserves, 15, 17, 19, 34, 58, 128
Rockefeller, 14
Ronald Reagan, 25
Royal Dutch, 14–15

S

S&P Goldman Sachs Commodity Index, 81
Saudi Arabia, 15, 17–19, 78
Savings, 47, 50–51, 73, 90, 95
Self-destructing, 46
Self-fulfilling, 45, 46, 104
Sell, 18, 25, 48, 50, 110, 112, 127
September 11th, 96
Seven Sisters, 1, 2, 15
Shah, 19
Shareholder value, 30
Shell, 14, 15
Shell Transport and Trade, 14
Short positions, 82, 98, 99, 121
Silver, 135
Social cleavages, 29
Social organisation, 32
Solar, 14
Soviet Union, 17
Soybeans, 135
Spare-capacity, 78, 99, 102, 129
Speculation, 48, 72, 98, 100, 102, 103, 110, 111, 112, 121, 123, 124, 129–131
Speculative investments, 121
Speculative investors, 131
Speculators, 34, 68, 100, 110, 112, 128, 129
Spill over, 35, 86, 122
Spot price, 66, 68, 102, 124, 125, 128, 131
Spread, 35, 46, 48, 73, 100, 103, 121, 131
Standard Bank, 118
Standard Oil, 14, 15
Standard Oil of California, 15
Standard Oil Company of New York, 15
Standard Oil of New Jersey, 15
State Department, 16, 17
Stock market(s), 42, 46, 48, 73, 80, 86, 103, 129
Stock prices, 42, 47, 48, 86
Stock returns, 43, 87, 130
Structural, 31
Subprime mortgages, 93
Suez Canal, 16
Supply shortage, 18, 129
Sustainability, 26, 87
Swap(s), 67, 68, 82, 89, 120, 131
Systemic, 26–28, 30–31, 34, 40
Systemic approaches, 27, 30

T

Tax, 15, 19, 71
Taxation, 87, 97
Technological advancement, 56, 131
Technological advances, 14, 56, 57, 59
Tehran, 16, 17
Texaco, 14, 15
Texas, 14, 113
Texas Corporation, 14
Three phases, 42, 60, 109, 112
Trade balance, 72
Traditional actors, 55, 59
Trans-Arabian oil pipeline, 16
Transnational corporations, 20
Transport, 15
Triangular structure, 20, 111, 124, 125
Tripoli, 16, 17

U

UK, 16, 20, 25, 73, 95, 98, 99, 112, 118, 120, 121, 125, 130, 131
UK Cabinet Office, 98, 99, 120, 125, 130
Uncertainty, 30

Unemployment, 71, 77, 79, 80, 86, 88, 94, 95, 103, 104
UN Trade and Development Report, 121
UPIA, 90
US, 2, 6, 15, 16, 17, 19, 20, 25, 77–80, 82, 86–89, 93–97, 100–104, 120–122, 124, 135, 141, 143–147
US dollar, 19, 25, 80, 82, 97, 100–102, 122
Utility Companies, 112

V
Venezuela, 15, 88, 101
Vietnam, 17, 18
Volatility, 30, 41, 52, 54, 55, 57, 67, 71, 82, 98, 110, 112, 122, 131, 132

W
Wall Street, 80, 95, 121, 131
Western Europe, 20
West Texas Intermediate, 67
Wheat, 135
Wind, 14, 87
WTI, 6, 67, 99, 141

The manufacturer's authorised representative in the EU is Springer Nature Customer Service Centre GmbH, Europaplatz 3, 69115 Heidelberg, Germany. If you have any concerns regarding our products, please contact ProductSafety@springernature.com

Printed and bound by CPI Group (UK) Ltd, Croydon, CR0 4YY

23/03/2026

02076402-0009